CLIMATE CHANGE AND ITS CAUSES, EFFECTS AND PREDICTION

FEDERAL CLIMATE CHANGE FUNDING

ANALYSES AND TRENDS

CLIMATE CHANGE AND ITS CAUSES, EFFECTS AND PREDICTION

Additional books in this series can be found on Nova's website under the Series tab.

Additional E-books in this series can be found on Nova's website under the E-book tab.

CLIMATE CHANGE AND ITS CAUSES, EFFECTS AND PREDICTION

FEDERAL CLIMATE CHANGE FUNDING

ANALYSES AND TRENDS

SANDRIN TOULLART
EDITOR

New York

For permission to use material from this book please contact us:
Telephone 631-231-7269; Fax 631-231-8175
Web Site: http://www.novapublishers.com

Library of Congress Cataloging-in-Publication Data

ISBN: 978-1-62948-554-6

Published by Nova Science Publishers, Inc. † New York

CONTENTS

Preface		**vii**
Chapter 1	Federal Climate Change Funding from FY2008 to FY2014 *Jane A. Leggett, Richard K. Lattanzio and Emily Bruner*	**1**
Chapter 2	Federal Climate Change Expenditures Report to Congress *Office of Management and Budget*	**21**
Chapter 3	The Global Climate Change Initiative (GCCI): Budget Authority and Request, FY2010-FY2014 *Richard K. Lattanzio*	**75**
Index		**97**

PREFACE

The federal government has funded work to address global climate change for more than four decades. An initial focus on science has expanded to encompass both mitigation and adaptation, involving at least 18 agencies plus the Executive Office of the President. The work supported is conducted by universities, national laboratories, private contractors, non-governmental organizations, and some federal agencies. Most of the funding has supported scientific and, since the 1990s, technological research and development. Given uncertainties regarding the risks of future climate change, the federal climate strategy has aimed at improving the information available for decision-making and reducing the costs of technologies that could help abate the risks. A growing component has been federal planning and efforts to adapt to climate change. Complementing the science and technology initiatives have been regulatory actions; programs to build capacity in private, state, local, and international entities to address climate change; and tax incentives to stimulate deployment of low greenhouse gas-emitting technologies. This book summarizes direct federal funding identified as climate change-related from FY2008 enacted funding through FY2013 and the FY2014 request. It reports the Administration's estimates of tax revenues not received due to energy tax provisions that may reduce GHG emissions and identifies the programs and funding levels, as well as some qualifications and observations on reporting of federal funding, and offers some issues that members of Congress may wish to consider in deliberating U.S. climate change strategies.

Chapter 1 – Direct federal funding to address global climate change totaled approximately $77 billion from FY2008 through FY2013. The large majority—more than 75%—has funded technology development and deployment, primarily through the Department of Energy (DOE). More than one-third of the identified funding was included in the American Recovery and Reinvestment Act of 2009 (P.L. 111-5). The President's request for FY2014 contains $11.6 billion for federal expenditures on programs. In the request, 23% would be for science, 68% for energy technology development and deployment, 8% for international assistance, and 1% for adapting to climate change. The Office of Management and Budget (OMB) also reports that energy tax provisions that may reduce greenhouse gas (GHG) emissions would reduce tax revenues by $9.8 billion.

At least 18 federal agencies administer climate change-related activities, according to OMB. Federal policy on climate change has been built largely from the "bottom up" from a variety of existing programs and mandates, presidential initiatives, and congressionally directed activities; funding has largely reflected departmental missions and support for each activity. Recently, the Obama Administration, in the context of its Climate Action Plan announced in June 2013, outlined an overall strategy with programs, resources, and tax incentives in a cross-agency, intergovernmental initiative. The new Climate Action Plan and a recent OMB report required by Congress on federal funding for climate change activities outline four main components of the strategy:

- Climate and Global Change Research and Education
- Reducing Emissions through Clean Energy Investments and Standards
- International Leadership
- Climate Change Adaptation

Possible Funding-Related Issues for Congress: Some Members of Congress have expressed interest in how federal funding may reflect and enable the Obama Administration's overall strategy, and priorities within it, to address climate change. Legislative issues regarding the federal funding of climate change activities may include the following:

- the sufficiency and alignment of federal resources to support a strategy to achieve long-term climate change policy goals;
- the demands of climate change adaptation programming for federal agencies, their programs, and resources;
- whether additional and predictable foreign aid resources may be provided to support actions by low-income countries to mitigate greenhouse gases or adapt to climate change;

- possible legislative proposals to restructure or improve collaboration among agencies regarding climate change activities;
- the incorporation of recommendations from evaluations (whether internal or external) to improve climate change programs; and
- possible requirements for reporting to Congress of funding, budget justifications, and programmatic progress that are adequate to support congressional decision-making and oversight.

Scope and Purpose of This Report: This report summarizes direct federal funding identified as climate change-related from FY2008 enacted funding through FY2013 and the FY2014 request (as well as a less consistent series beginning with FY2001). It reports the Administration's estimates of tax revenues not received due to energy tax provisions that may reduce GHG emissions. The report briefly identifies the programs and funding levels, as well as some qualifications and observations on reporting of federal funding. It further offers some issues that Members may wish to consider in deliberating on U.S. climate change strategies.

Chapter 2 – The U.S. Government's portfolio of climate change programs and cross-cutting initiatives focuses on advancing our understanding of climate change and its impact on our communities; advancing the development and introduction of energy-efficient, renewable, and other low- or non-emitting technologies; improving standards for measuring and registering emissions reductions and supporting preparedness and resilience to climate change impacts. Many elements of the Administration's climate change portfolio are designed to provide incentives for greenhouse gas (GHG) emissions reductions domestically to support community-based preparedness and resilience efforts, to ensure that Federal operations and facilities continue to protect and serve citizens in a changing climate, and to promote international initiatives focused on concrete actions toward reducing greenhouse gas emission and enhance climate preparedness globally.

Chapter 3 – The United States supports international financial assistance for global climate change initiatives in developing countries. Under the Obama Administration, this assistance has been articulated primarily as the Global Climate Change Initiative (GCCI), a platform within the President's 2010 Policy Directive on Global Development. The GCCI aims to integrate climate change considerations into U.S. foreign assistance through a range of bilateral, multilateral, and private sector mechanisms to promote sustainable and climate-resilient societies, foster low-carbon growth, and reduce emissions from deforestation and land degradation. The GCCI is implemented through programs at three "core" agencies: the Department of State, the Department of the Treasury, and the U.S. Agency for International Development (USAID). Most GCCI activities at USAID are implemented through the agency's bilateral development assistance programs. Many of the GCCI activities at the Department of State and the Department of the Treasury are implemented through international organizations, including the United Nations Framework Convention on Climate Change's Least Developed Country Fund and Special Climate Change Fund, as well as multilateral financial institutions such as the Global Environment Facility, the Clean Technology Fund, and the Strategic Climate Fund. The GCCI is funded through the Administration's Executive Budget, Function 150 account, for State, Foreign Operations, and Related Programs.

Congress is responsible for several activities in regard to the GCCI, including (1) authorizing periodic appropriations for federal agency programs and multilateral fund contributions, (2) enacting those appropriations, (3) providing guidance to the agencies, and (4) overseeing U.S. interests in the programs and the multilateral funds. Recent budget authority for the GCCI was $323 million in FY2009, $945 million in FY2010, $819 million in FY2011, and $858 million in FY2012, and has been enacted through legislation including the Omnibus Appropriations Act, 2009 (H.R. 1105; P.L. 111-8); the Consolidated Appropriations Act, 2010 (H.R. 3288; P.L. 111-117); the Supplemental Appropriations Act, 2010 (H.R. 4899; P.L. 111-212); the Department of Defense and Full-Year Continuing Appropriations Act, 2011 (H.R. 1473; P.L. 112-10); and the Consolidated Appropriations Act, 2012 (H.R. 2055; P.L. 112-74). FY2013 contributions to GCCI programming as provided for in the Consolidated and Further Continuing Appropriations Act, 2013 (H.R. 933; P.L. 113-6), have yet to be fully reported by the agencies. The Administration's FY2014 GCCI budget request is $837 million. Congressional committees of jurisdiction for the GCCI include the U.S. House of Representatives Committees on Foreign Affairs (various subcommittees); Financial Services, Subcommittee on Monetary Policy and Trade; and Appropriations, Subcommittee on State, Foreign Operations, and Related Programs; and the U.S. Senate Committees on Foreign Relations, Subcommittee on International Development and Foreign Assistance, Economic Affairs, and International Environmental Protection; and Appropriations, Subcommittee on State, Foreign Operations, and Related Programs.

As Congress considers potential authorizations and/or appropriations for activities administered through the GCCI, it may have questions concerning

U.S. agency initiatives and current bilateral and multilateral programs that address global climate change. Some potential concerns may include cost, purpose, direction, efficiency, and effectiveness, as well as the GCCI's relationship to industry, investment, humanitarian efforts, national security, and international leadership. This report serves as a brief overview of the GCCI and its structure, intents, and funding history.

In: Federal Climate Change Funding
Editor: Sandrin Toullart
ISBN: 978-1-62948-554-6

Chapter 1

FEDERAL CLIMATE CHANGE FUNDING FROM FY2008 TO FY2014*

Jane A. Leggett, Richard K. Lattanzio and Emily Bruner

SUMMARY

Direct federal funding to address global climate change totaled approximately $77 billion from FY2008 through FY2013. The large majority—more than 75%—has funded technology development and deployment, primarily through the Department of Energy (DOE). More than one-third of the identified funding was included in the American Recovery and Reinvestment Act of 2009 (P.L. 111-5). The President's request for FY2014 contains $11.6 billion for federal expenditures on programs. In the request, 23% would be for science, 68% for energy technology development and deployment, 8% for international assistance, and 1% for adapting to climate change. The Office of Management and Budget (OMB) also reports that energy tax provisions that may reduce greenhouse gas (GHG) emissions would reduce tax revenues by $9.8 billion.

At least 18 federal agencies administer climate change-related activities, according to OMB. Federal policy on climate change has been

* This is an edited, reformatted and augmented version of a Congressional Research Service publication, CRS Report for Congress R43227, prepared for Members and Committees of Congress, from www.crs.gov, dated September 13, 2013.

built largely from the "bottom up" from a variety of existing programs and mandates, presidential initiatives, and congressionally directed activities; funding has largely reflected departmental missions and support for each activity. Recently, the Obama Administration, in the context of its Climate Action Plan announced in June 2013, outlined an overall strategy with programs, resources, and tax incentives in a cross-agency, intergovernmental initiative. The new Climate Action Plan and a recent OMB report required by Congress on federal funding for climate change activities outline four main components of the strategy:

- Climate and Global Change Research and Education
- Reducing Emissions through Clean Energy Investments and Standards
- International Leadership
- Climate Change Adaptation

Possible Funding-Related Issues for Congress

Some Members of Congress have expressed interest in how federal funding may reflect and enable the Obama Administration's overall strategy, and priorities within it, to address climate change. Legislative issues regarding the federal funding of climate change activities may include the following:

- the sufficiency and alignment of federal resources to support a strategy to achieve long-term climate change policy goals;
- the demands of climate change adaptation programming for federal agencies, their programs, and resources;
- whether additional and predictable foreign aid resources may be provided to support actions by low-income countries to mitigate greenhouse gases or adapt to climate change;
- possible legislative proposals to restructure or improve collaboration among agencies regarding climate change activities;
- the incorporation of recommendations from evaluations (whether internal or external) to improve climate change programs; and
- possible requirements for reporting to Congress of funding, budget justifications, and programmatic progress that are adequate to support congressional decision-making and oversight.

Scope and Purpose of This Report

This report summarizes direct federal funding identified as climate change-related from FY2008 enacted funding through FY2013 and the

FY2014 request (as well as a less consistent series beginning with FY2001). It reports the Administration's estimates of tax revenues not received due to energy tax provisions that may reduce GHG emissions. The report briefly identifies the programs and funding levels, as well as some qualifications and observations on reporting of federal funding. It further offers some issues that Members may wish to consider in deliberating on U.S. climate change strategies.

INTRODUCTION

The federal government has funded work to address global climate change for more than four decades. An initial focus on science has expanded to encompass both mitigation and adaptation, involving at least 18 agencies plus the Executive Office of the President. The work supported is conducted by universities, national laboratories, private contractors, non-governmental organizations, and some federal agencies. Most of the funding has supported scientific and, since the 1990s, technological research and development (R&D).

Given uncertainties regarding the risks of future climate change, the federal climate strategy has aimed at improving the information available for decision-making and reducing the costs of technologies that could help abate the risks. A growing component has been federal planning and efforts to adapt to climate change. Complementing the science and technology initiatives have been regulatory actions;[1] programs to build capacity in private, state, local, and international entities to address climate change; and tax incentives to stimulate deployment of low greenhouse gas-emitting technologies. Many of these initiatives are identified in the Obama Administration's recently announced Climate Action Plan, which stated three main prongs:[2]

- cutting carbon pollution in America;
- preparing the United States for the impacts of climate change; and
- leading international efforts to address global climate change.

Federal funding for these activities described in this report differs across these priorities and has shifted over time. Policy instruments and programs depend on funding to differing degrees to be effective; some rely on large amounts of direct federal investment (e.g., in federal R&D) while others primarily require support for administrative expenses.[3]

As congressional and public debate continues with regard to whether and how to address climate change, the priorities for federal funding are likely to evolve further. Members of Congress have expressed a range of views about funding for climate change-related activities. Some question the relative priorities among initiatives or whether the risks of climate change merit the magnitude of federal expenditures and of federal policies on the economy in exchange for benefits that would mostly accrue to future generations, people in other countries, and stability of Earth systems. Other Members point to scientific and economic research to underpin their support for increasing funding to address climate change. Congressional debate on climate change funding takes place in the broader context of stark choices among competing fiscal demands.

FEDERAL FUNDING FOR CLIMATE CHANGE

The Office of Management and Budget (OMB) and federal agencies have identified approximately $77 billion of budget authority available to federal agencies from Fiscal Years 2008 through 2013 for climate change activities. The large majority—more than 75%—funded technology development and deployment, mostly through the Department of Energy (DOE). More than one-third of the identified funding during this period was appropriated in the American Recovery and Reinvestment Act of 2009 (P.L. 111-5), enacted February 17, 2009.

This CRS report presents information available on federal budget authority in FY2008 through FY2013, and the President's FY2014 request, for climate change activities of federal departments and agencies. The amounts are reported by OMB, and occasionally by individual departments and agencies where noted. CRS relies on OMB and agency sources of data since most climate-related funding is appropriated at the subaccount level and therefore is not directly identifiable in legislation and committee reports. OMB has produced a "budget cross-cut" for climate change activities when it has been required by language in appropriations bills.[4] Individual agencies often report funding for specific programs, but that information may not be comprehensive or comparable to information on funding for climate change activities in other programs or agencies. Further, when agencies cooperate on programs, double-counting of funding could occur if one were simply to add up what the agencies each report.[5] Some of the OMB and agency sources may report inconsistent or incomplete data.[6] It is CRS's judgment that the amounts in the following tables likely underestimate total federal funding relating to climate change for the period, perhaps on the order of tens of millions of dollars (i.e., not billions). Information is not available for all programs for all years, as explained below.

Climate Change Initiatives

Tables in this report detail funding by agency in terms of budget authority.[7] Funding information is provided for several of the major federal initiatives which address climate change, including the following:

- the Global Change Research Program;
- Clean Energy Technologies, largely corresponding to the former Climate Change Technology Program;
- International Assistance, sometimes called the Global Climate Change Initiative; and
- Climate Adaptation, Preparedness, and Resilience.

The President's request for FY2014 contains $11.6 billion for federal expenditures on programs. In the request, 23% would be for science (the U.S. Global Change Research Program or USGCRP), 68% for "clean energy" technology development and deployment, 8% for international assistance, and 1% for adapting to climate change. The Office of Management and Budget (OMB) also reports energy tax provisions that may reduce greenhouse gas (GHG) emissions would reduce tax revenues by $9.8 billion.

The Office of Management and Budget (OMB) also reports estimates of Energy Tax Provisions and Energy Grants in Lieu of Tax Provisions that may stimulate deployment of low GHGemitting technologies. These fiscal incentives serve as exemptions to the baseline tax structure, and usually result in a reduction in the amount of tax owed. A review of these provisions is beyond the scope of this CRS report, though the OMB estimates are provided in *Table 4*.[8]

Minor inconsistencies sometimes exist in alternative reports on funding, perhaps due to rescissions or reprogramming, or because a program may be cast as climate change-related in some contexts but not in others. The reported initiatives are cross-agency "roll-ups" of programs and funding in the agencies, and information on some programs is available only to the degree that the agency has reported funding to OMB or Congress.

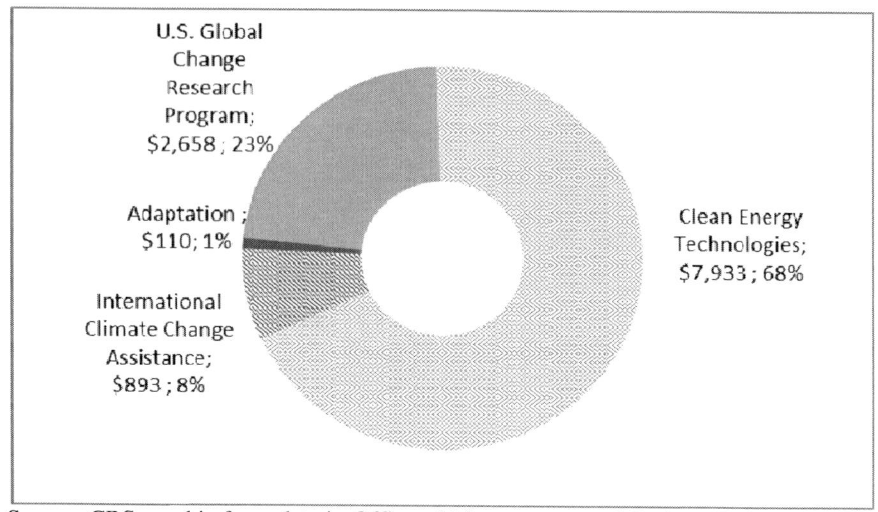

Source: CRS graphic from data in Office of Management and Budget. Federal Climate Change Expenditures Report to Congress. Washington, D.C: Executive Office of the President, August 2013.

Figure 1. The President's Request for Climate Change-Related Budget Authority for FY2014. (in millions of nominal 2013 dollars and percent of the request; excludes tax provisions).

Some activities, particularly those that consider or address potential impacts of climate change on federal programs or assets, likely existed prior to their being identified as climate change-related, so that funding may have occurred prior or in addition to activities listed in these tables.

This CRS report also provides a series of reported enacted budget authority and tax provisions for climate change programs since FY2001, in *Table 4.* One may presume that past numbers are increasingly inconsistent with present accounting the farther back in time, for reasons explained above. Nonetheless, the general magnitude of overall spending and the shifts in categories of spending may be useful to some. *Figure 2* shows those estimates of historical budget authority since FY2001 to the FY2014 request, adjusted for inflation (i.e., in constant 2012 dollars).

Through most of the 1990s, most federal funding identified to address climate change was directed at improving the science, under the U.S. GCRP.[9] As *Figure 2* shows, by FY2001, investment in climate-related energy technology research, development, and deployment (Clean Energy Technologies) surpassed science (the U.S. GCRP) as the largest component of federal activities and has continued to increase its share since then; the share in FY2013 enacted budget authority shrunk slightly from FY2012, however.

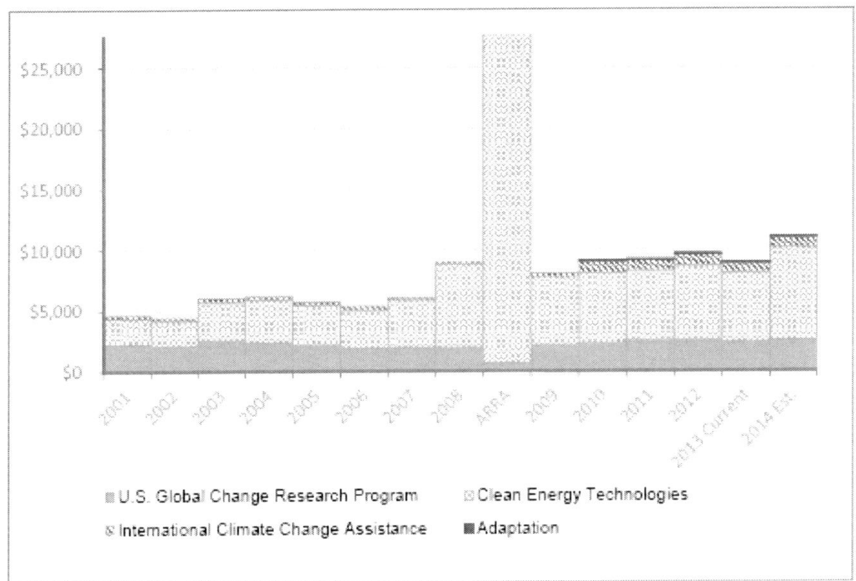

Source: CRS figure using the White House Office of Management and Budget. Federal Climate Change Expenditures Report to Congress. Washington, D.C: Executive Office of the President, August 2013, as well as past OMB reports released in 2002, 2003, 2004, 2005, 2006, 2008, and 2010;

Notes: To inflate historical figures in 2012 dollars, CRS used OMB, The Budget for Fiscal Year 2014, Historical Tables, Table 10.1- Gross Domestic Product and Deflators Used in the Historical Tables: 1940-2018. CRS used the index for non-defense expenditures and adjusted the index so FY2012=1.0.

Names for groupings of programs aimed generally at the same goals have shifted through successive Administrations. The U.S. Global Change Research Program (USGCRP) was supplemented by the early 2000s with the Climate Change Science Program (CCSP), which is now folded into the USGCRP. Similarly, energy development and deployment programs, called the Climate Technologies Initiative in the late 1990s and the Climate Change Technologies Program (CCTP) through the 2000s, is largely consistent with the Clean Energy Technologies category reported most recently by OMB and illustrated in this figure. This time series presents the totals reported without attempting to adjust for minor changes in scope and emphasis across the initiatives.

Figure 2. Federal Budget Authority for Climate Change Programs, FY2001 to the FY2014 Request, by Major Category of Program (in millions of constant 2012 dollars).

The one-time investment in climate-related technologies contained in the American Recovery and Reinvestment Act of 2009 (P.L. 111-5) was notably large and still constitutes more than one-fifth of all identified federal expenditures since FY2001 (including the FY2014 request).

Adaptation to climate change has been added to the identified federal expenditures since FY2010, but remains too small to be seen as a share on the time series figure.

The Global Change Research Program (USGCRP)

The U.S. Global Change Research Program (USGCRP) was mandated by Congress in the Global Change Research Act of 1990 (P.L. 101-606). The USGCRP is intended to improve understanding of climate science, including the cumulative effects of human activities and natural processes on the environment; develop science-based resources to support policymaking and resource management; and communicate findings broadly among scientific and stakeholder communities. Thirteen departments and agencies participate in the USGCRP.[10] The White House Office of Science and Technology Policy (OSTP) and the Office of Management and Budget (OMB) work with the USGCRP to establish research priorities and funding plans to help ensure that the program is aligned with the Administration's priorities and reflects agency planning.

Table 1 below presents estimates for direct federal funding for the USGCRP from FY2008 through the FY2014 request. During the George W. Bush Administration, a second set of scientific research programs was identified and named the Climate Change Science Program. The CCSP is now merged into the USGCRP in the cross-cut budget.

Clean Energy Technologies

The Clean Energy Technologies (CET) efforts are intended to research, develop, and deploy technologies that would reduce GHG emissions compared to technologies currently commonly used. Besides federal R&D, CET includes a variety of voluntary partnership and grant activities.

Table 1. Budget Authority for the Global Change Research Program (USGCRP) (in millions of nominal dollars)

Agency	2008 Actual	ARRA Enacted	2009 Enacted	2010 Enacted	2011 Enacted	2012 Enacted	2013 Current[a]	2014 Request
Department of Agriculture	$63	$0	$0	$121	$116	$116	$106	$126
Department of Commerce	272	218	377	363	338	319	302	371
Department of Energy	128	65	168	171	183	211	209	220
Department of Health and Human Services	4	0	5	4	4	14	14	15
Department of the Interior	34	0	45	63	64	59	55	72
Department of State	0	0	0	14	3	3	3	3
U.S. Agency for International Development	0	0	0	28	25	11	11	14
Department of Transportation	1	0	2	3	1	1	1	1
Department of the Treasury	0	0	0	0	0	0	0	0
Environmental Protection Agency	17	0	18	21	20	18	17	20
National Aeronautics and Space Administration	1,084	237	1,086	1,123	1,431	1,427	1,435	1,499
National Science Foundation	207	121	269	320	321	333	316	326
Smithsonian Institution	6	0	6	7	7	8	8	8
Totals	$1,816	$641	$2,023	$2,238	$2,513	$2,506	$2,463	$2,658

Source: Office of Management and Budget. Federal Climate Change Expenditures Report to Congress. Washington, DC: Executive Office of the President, June 10, 2010, and August 28 2013; Office of Management and Budget. Analytical Perspectives—Budget of the U.S. Government. Washington DC: Executive Office of the President, 2010, 2011, 2012, and 2013. USGCRP website, http://www.globalchange.gov/about/budget-documents. For FY2012 and FY2014, http://www.whitehouse.gov/sites/default/files/microsites/ostp/2014_R&Dbudget_climate

Notes: N.A., or "not available," means that CRS could not identify appropriate figures.

Regular year funding for FY2009 and supplemental appropriations in the American Recovery and Reinvestment Act of 2009 (ARRA) (P.L. 111-5) are identified separately in this table.

[a] FY2013 Current Budget Authority, according to OMB, as of June 21, 2013. It reflects funds available calculated as the amount appropriated (FY2013 Enacted Budget Authority, data not shown), minus reductions triggered by the Budget Control Act of 2011 (P.L. 112-25) sequestration order issued March 1, 2013.

Funding provided by the CET initiative is intended to stimulate the development and use of selected energy technologies, including renewable, low-carbon fossil, and nuclear technologies as well as energy efficient technologies, products, and process improvements. Consequently, the large majority of CET budget authority has been in the Department of Energy.

The CET initiative is a recent iteration of the Climate Change Technology Program (CCTP) that was formally established administratively under President George W. Bush, which itself included a number of existing programs. The FY2014 budget request calls for an investment of nearly $7.9 billion in support of CET, a 30% and 37% increase in comparison to the enacted FY2012 and current FY2013 budget authority, respectively. These funds reflect the current Administration's priority for new technologies as a means to facilitate future GHG reductions and meet the President's GHG emissions reduction targets, in the range of 17% below 2005 levels by 2020 and approximately 83% below 2005 levels by 2050. Meeting the 2050 target would likely require radical change from the currently prevalent energy system.

The Department of Energy coordinated the CCTP inter-agency effort through a central office. It is unclear to what degree the current programs are coordinated as an inter-agency strategy, or through what mechanism. However, as is apparent in the budget figures below, most of the activities are within the Department of Energy.

Table 2 presents estimates, to the extent they are available, for direct federal funding for the CCTP from FY2008 through the FY2014 request.

Table 2. Budget Authority for Clean Energy Technologies (Largely the Former Climate Change Technology Program) (in millions of nominal dollars)

Agency	2008 Actual	ARRA Enacted	2009 Enacted	2010 Enacted	2011 Enacted	2012 Enacted	2013 Current[a]	2014 Request
Department of Agriculture	$205	$271	$100	$453	N.A.	$275	$305	$265
Department of Commerce	8	4	15	18	N.A.	40	40	43
Department of Defense	150	139	261	226	N.A.	481	437	457

Agency	2008 Actual	ARRA Enacted	2009 Enacted	2010 Enacted	2011 Enacted	2012 Enacted	2013 Current[a]	2014 Request
Department of Energy	5,745	25,223	4,543	4,399	5,503	4,388	4,144	6,212
Department of Transportation	19	100	43	125	N.A.	91	56	52
Environmental Protection Agency	107	0	111	133	113	117	111	115
National Aeronautics and Space Administration	139	31	119	124	N.A.	296	276	321
National Science Foundation	21	2	24	26	30	341	346	372
Nuclear Regulatory Commission	N.A.	N.A.	N.A.	N.A.	N.A.	83	57	86
Tennessee Valley Authority	N.A.	N.A.	N.A.	N.A.	N.A.	9	11	10
Totals	$6,394	$25,770	$5,216	$5,504	$5,646	$6,121	$5,783	$7,933

Source: Climate Change Technology Program, workbook of detailed funding data provided to CRS, 2008; Office of Management and Budget. Federal Climate Change Expenditures Report to Congress. Washington, DC: Executive Office of the President, June 10, 2010, and August 28, 2013; DOE CCTI estimates in FY2010-2012 enacted also from http://www.cfo.doe.gov/budget/12budget/Content/Volume2.pdf (p. 200); and (p. 135); NSF enacted for CCTP also from http://nsf.gov/about/budget/fy2012/pdf/fy2012_rollup.pdf, p. 12, and http://www.nsf.gov/about/budget/fy2013/pdf/02-Summary_Tables_fy2013.pdf.

Notes: N.A., or "not available," means that CRS could not identify appropriate figures.

Regular year funding for FY2009 and supplemental appropriations in the American Recovery and Reinvestment Act of 2009 (ARRA) (P.L. 111-5) are identified separately in this table.

[a] FY2013 Current Budget Authority, according to OMB, as of June 21, 2013. It reflects funds available calculated as the amount appropriated (FY2013 Enacted Budget Authority, data not shown), minus reductions triggered by the Budget Control Act of 2011 (P.L. 112-25) sequestration order issued March 1, 2013.

International Assistance

The United States supports international financial assistance for global climate change initiatives in developing countries. Under the Obama Administration, this assistance has been articulated primarily as the Global Climate Change Initiative (GCCI), a platform within the President's 2010 Policy Directive on Global Development. The GCCI aims to integrate climate change considerations into U.S. foreign assistance through a range of bilateral, multilateral, and private sector mechanisms to promote sustainable and climate-resilient societies, foster low-carbon growth, and reduce emissions from deforestation and land degradation. The GCCI is implemented primarily through programs at three "core" agencies: the Department of State, the Department of the Treasury, and the U.S. Agency for International Development (USAID). Most GCCI activities at USAID are implemented through the agency's bilateral development assistance programs. Many of the GCCI activities at the Department of State and the Department of the Treasury are implemented through international organizations, including the United Nations Framework Convention on Climate Change's Least Developed Country Fund and Special Climate Change Fund, as well as multilateral financial institutions such as the Global Environment Facility, the Clean Technology Fund, and the Strategic Climate Fund.

Table 3 presents estimates for federal budget authority for the GCCI from FY2008 through the FY2014 request.[11]

Table 3. Budget Authority for the International Climate Change Assistance (Also Referred to as the Global Climate Change Initiative, or GCCI) (in millions of nominal dollars)

Agency	2008 Actual	ARRA Enacted	2009 Enacted	2010 Enacted	2011 Enacted	2012 Enacted	2013 Current[a]	2014 Request
Department of Agriculture	$0	$0	$4	$5	N.A.	$3	$3	$1
Department of Commerce	0	0	11	11	N.A.	N.A.	N.A.	N.A.
Department of Energy	0	0	0	13	N.A.	13	13	13
Department of State	41	0	55	202	125	133	126	133
US Agency for International Development	115	0	222	306	398	348	335	349
Department of the Treasury	46	0	46	438	296	377	296	356
Environmental Protection Agency	0	0	20	21	N.A.	18	16	19
Millennium Challenge Corporation	0	0	0	2	N.A.	41	0	0
National Aeronautics and Space Administration	0	0	2	2	N.A.	3	3	3
National Science Foundation	0	0	3	6	N.A.	6	6	3
US Trade and Development Agency	0	0	10	17	N.A.	16	0	18
Totals	$202	$0	$373	$1,023	$819	$958	$797	$893

Source: Office of Management and Budget. Federal Climate Change Expenditures Report to Congress. Washington,

DC: Executive Office of the President, June 10, 2010, and August 28, 2013; CRS Report R41845, The Global Climate Change Initiative (GCCI): Budget Authority and Request, FY2010-FY2014, by Richard K. Lattanzio.

Notes: N.A., or "not available," means that CRS could not identify appropriate figures.

Regular year funding for FY2009 and supplemental appropriations in the American Recovery and Reinvestment Act of 2009 (ARRA) (P.L. 111-5) are identified separately in this table.

OMB began to provide a budget cross-cut for the GCCI for FY2011, identifying specific funding amounts for the principal international agencies involved and noting that other agencies contribute. OMB did not identify specific enacted budget authorities for those other agencies, and doing so would likely double-count at least some of the funds represented in other tables. Where CRS understands that no funds were available, a zero is used; in those cases where some funds may be used in conjunction with the GCCI, CRS uses N.A. in this table.

[a] FY2013 Current Budget Authority, according to OMB, as of June 21, 2013. It reflects funds available calculated as the amount appropriated (FY2013 Enacted Budget Authority, data not shown), minus reductions triggered by the Budget Control Act of 2011 (P.L. 112-25) sequestration order issued March 1, 2013.

CLIMATE ADAPTATION, PREPAREDNESS, AND RESILIENCE

Funding does not appear to be reported by most agencies in support of their plans to adapt to climate change, or "Adaptation Plans," as stipulated under Executive Order 13514 and reported since 2012 in agencies' Strategic Sustainability Performance Plans (SSPPs). Many agencies are in early stages of planning or implementation, and the related funding would in almost all cases be in the details of appropriations accounts, and therefore not readily available.

One exception is the Department of the Interior (DOI). The Secretary of Interior issued Secretarial Order 3289 in 2009 (amended in 2010), "addressing the Impacts of Climate Change on America's Water, Land, and Other Natural and Cultural Resources." DOI has identified vulnerabilities to its missions and assets, as well as to tribes and their resources, and has developed plans and associated budgets to address those vulnerabilities. According to OMB, enacted budget authority for this purpose increased from $88 million in FY2012 to $90 million in FY2013 (after sequestration). The request for FY2014 was $110 million. These amounts do not include resources in the U.S.

Geological Survey, as their efforts to provide related science to support the bureaus and other agencies are reported in the USGCRP.

PRINCIPAL OBSERVATIONS AND LEGISLATIVE ISSUES

Over the past two decades, federal funding for climate change-related activities has expanded from scientific research, almost exclusively, to a wide variety of programs to (1) develop and disseminate technologies; (2) build an informational and analytical foundation for future policy actions; (3) plan for adaptation to actual or expected climate change; (4) assist private sector decision-makers and lower-income countries; and (5) address additional needs. Federal strategy on climate change largely reflects the aggregation of a variety of existing programs, presidential initiatives, and congressionally directed activities.

Presidential initiatives have been built largely from existing programs, and agencies have contributed to shared program goals using existing resources and expertise. Priorities tend to be based on departmental missions and the degree of support for the input activities. As the debate continues over appropriate strategies to address climate change, the needs and priorities for funding are likely to evolve further.

Federal budget for climate change-related programs is primarily aimed at investment and tax provisions aimed at stimulating energy technologies that may reduce GHG emissions. The Clean Energy Technologies category represents 68% of the President's request for FY2014. Over the period since FY2001, including the FY2014 request, these technology programs constitute more than 70% of all funding for identified climate change-related programs (excluding tax provisions). By itself, the one-time injection of resources in the American Recovery and Reinvestment Act of 2009 (P.L. 111-5)—almost entirely for energy technology development—is nearly one-quarter the cumulative spending from FY2001 through FY2013.

Since FY2001, funding for climate change science has remained roughly constant, varying up or down by as much as 15% in real dollars (i.e., adjusted for inflation).

Funding for science was near its real-dollar low in FY2008, and would almost reach a high if the FY2014 request were enacted. (The peak was $2,577 million in FY2003 in 2012 constant dollars.)

International assistance, to facilitate capacity-building and efforts by selected, low-income countries to mitigate their GHG emissions and adapt to expected climate change, has nearly tripled from FY2001 to the current FY2013 budget authority. The FY2014 request proposed an increase of the current FY2013 amount by 9.5%, to $856 million. This is 8% of the President's request.

Adaptation to climate change has grown in attention across the federal agencies, as well as across segments of the public and private sector. Although activities in the federal agencies have expanded, especially since 2009, little of this activity is reported in OMB's budget cross-cut. This omission probably reflects the relatively low level of effort in most agencies, and that efforts are typically add-ins to wider programs rather than stand-alone activities. For example, grants for coastal zone planning may incorporate effects of potential sea level rise and storm surges as one consideration among many in those plans. Only the Department of the Interior, which has a department-wide effort to identify and address vulnerabilities to the assets it manages, explicitly reports adaptation funds. This is less than 1% of the FY2013 funding across agencies.

Interpreting how funding relates to levels of effort to address climate change is challenged by several reporting issues. Identifying actual funding for climate change activities is clouded by several ongoing issues, including the following:

- the levels of aggregation of the budget request;
- changes in scope of what is reported;
- changes in accounting methods used over time;
- inconsistencies across agencies in defining and interpreting methods for reporting activities;
- lack of description by agencies for subaccount level climate change-related activities in their budget documentation; and
- omissions of reporting of some arguably climate-related activities in the overall program.

In light of these issues, the Government Accountability Office (GAO) investigated the Administration's reporting practices in 2005, 2006, and 2011.[12] GAO recommended that the Administration provide greater clarity and consistency of reporting on federal funding for climate change activities. While some improvements were made following the 2006 GAO report, many issues persist, confounding analysis of climate change funding. The 2011

GAO report concluded that there were two key factors that "complicate efforts to align funding with priorities": (1) "... federal officials do not have a shared understanding of strategic priorities" and (2) "... since mechanisms for aligning funding with priorities are nonbinding, they are limited when in conflict with agencies' own priorities."[13] In addition, congressional priorities may differ from the President's goals and requests, further complicating alignment of goals and priorities.

The funding data described in this report derive primarily from OMB's series of *Reports to Congress on Federal Climate Change Expenditures* published through the requirements in various appropriation laws passed from 2002 through 2008, 2010, and 2012. In response to language included in FY2012 appropriations, OMB released its latest report on August 28, 2013. It appears that OMB produces these budget cross-cuts only when required by legislation.

If the 113th Congress finds data on climate change funding and programs useful for oversight and other legislative purposes, it may consider renewing requirements for reporting of climate change-related activities, including directions that would continue to improve clarity and consistency compared to previous reports.

Viewing Efforts Across Federal Agencies and Programs Through a Cross-Cut Budget

Compilation and strategic management of resource allocation and use across agencies for a particular policy, called a "cross-cut budget," is occasionally used in federal budgeting and appropriations.[14] On the one hand, cross-cut budgeting may be a useful exercise where a policy area or one or more policy goals engage a variety of departments and agencies. With a cross-cut budget, Congress, agencies, and non-federal stakeholders may seek to better understand and align resources and implementation across agencies and appropriations subcommittees. On the other hand, the development of cross-cut budgets also may add workload for related agencies and the Office of Management and Budget. For reasons stated in this report by GAO, and others, budget cross-cuts and long time-series may contain inconsistencies—due to changing terminology, concepts, and priorities—that track ongoing developments but also may limit the reliability and utility of underlying data, to some extent.

Table 4. Federal Climate Change Programs: FY2001-FY2013 Enacted, and the President's FY2014 Request (in millions of nominal dollars, except where noted)

Major Climate Change Program Areas	Budget Authority (Fiscal Years, Enacted Except FY2013 and FY2014)														
	2001	2002	2003	2004	2005	2006	2007	2008	ARRA	2009	2010	2011	2012	2013 Current[a]	*2014 Request*
U.S. Global Change Research Program (including the former Climate Change Science Program)	$1,728	$1,667	$2,078	$1,996	$1,864	$1,691	$1,825	$1,864	$641	$2,023	$2,195	$2,448	$2,506	$2,463	$2,658
Clean Energy Technologies (formerly Climate Change Technology Program)	$1,675	$1,637	$2,533	$2,870	$2,808	$2,789	$3,485	$6,394	$25,499	$5,216	$5,504	$5,646	$6,121	$5,783	$7,933
International Climate Change Assistance	$218	$224	$270	$252	$234	$249	$188	$202	$0	$323	$939	$819	$958	$797	$893
Adaptation	$0	$0	$0	$0	$0	$0	$0	$0	$0	$0	$69	$35a	$88	$90	$110
All Areas	$3,621	$3,528	$4,881	$5,118	$4,906	$4,729	$5,498	$8,460	$26,140	$7,562	$8,707	$8,948	$9,673	$9,133	*$11,594*
All Areas in Constant 2012 Dollars	*$4,650*	*$4,471*	*$6,053*	*$6,189*	*$5,745*	*$5,361*	*$6,076*	*$9,027*	*$27,739*	*$8,025*	*$9,112*	*$9,166*	*$9,673*	*$8,955*	*$11,112*
	Estimated Revenue Loss Effects of Energy Tax Provisions														
Energy Tax Provisions That May Reduce	2001	2002	2003	2004	2005	2006	2007	2008	ARRA	2009	2010	2011	2012	2013Est.	2014Est.
Greenhouse Gases	$0	$0	$580	$500	$369	$1,160	$1,520	$1,520	$0	$1,440	$4,140	N.A.	$10,132	$13,079	$9,839
Tax Expenditures in Constant 2012 dollars	$0	$0	$719	$605	$432	$1,315	$1,680	$1,622	$0	$1,528	$4,332	$0	$10,132	$12,824	$9,430
Total Budgetary Impact[b]	$3,621	$3,522	$5,461	$5,618	$5,275	$5,889	$7,018	$9,980	$26,140	$9,002	$12,277	N.A.	$19,805	$22,212	$21,433

Table 4. (Continued)

Major Climate Change Program Areas	Budget Authority (Fiscal Years, Enacted Except FY2013 and FY2014)														
	2001	2002	2003	2004	2005	2006	2007	2008	ARRA	2009	2010	2011	2012	2013 Current[a]	*2014 Request*
	Estimated Revenue Loss Effects of Energy Tax Provisions														
Total Budgetary Impact in Constant 2007 Dollars	$4,650	$4,471	$6,773	$6,793	$6,178	$6,676	$7,756	$10,648	$27,739	$9,553	$13,444	$9,166	$19,805	$21,779	$20,542

Source: Prepared by CRS based on data reported by OMB, including in Federal Climate Change Expenditures Report to Congress FY2014, August 2013; Federal Climate Change Expenditures Report to Congress, FY2008, May 2007, Table 8, p. 27; Federal Climate Change Expenditures Report to Congress, August 2003, Table 1; and Federal Climate Change Expenditures Report to Congress, July 2002, Table 1.

Notes: OMB adjusted the science and technology amounts for FY2003, FY2004, and FY2005 to reflect more recent accounting within these program areas. Remaining inconsistencies are likely across years due to changes in the scopes of what is considered for "climate change" and methods of allocating certain costs to programs.

To convert from nominal to 2012 dollars, CRS used OMB, The Budget for Fiscal Year 2014, Historical Tables, Table 10.1- Gross Domestic Product and Deflators Used in the Historical Tables: 1940-2018. CRS used deflators for non-defense expenditures and adjusted index so FY2012=1.0.

The total impact to the federal budget in each year as reported by OMB may differ somewhat from the sum of budget authority for each program area and the estimates of tax expenditures, due to exclusion by OMB of dollar amounts for certain activities to avoid double-counting.

[a] FY2013 Current Budget Authority, according to OMB, as of June 21, 2013. It reflects funds available calculated as the amount appropriated (FY2013 Enacted Budget Authority, data not shown), minus reductions triggered by the Budget Control Act of 2011 (P.L. 112-25) sequestration order issued March 1, 2013.

[b] Because the estimate for the adaptation efforts in this year is incomplete, this figure is likely an underestimate of the actual amount.

Some Members of Congress have expressed interest in how federal funding may reflect and enable the Obama Administration's overall strategy, and priorities within it, to address climate change. Beyond the amounts of budget authority in this report, further legislative considerations regarding the federal funding of climate change activities may include the following:

- the sufficiency and alignment of federal resources to support a strategy to achieve long-term climate change policy goals;
- the demands of climate change adaptation programming for federal agencies, their programs, and resources;
- whether additional and predictable foreign aid resources may be provided to support actions by low-income countries to mitigate greenhouse gases or adapt to climate change;
- possible legislative proposals to restructure or improve collaboration among agencies regarding climate change activities; and
- the incorporation of recommendations from evaluations (whether internal or external) to improve climate change programs.

End Notes

[1] Regulatory actions have included some explicitly required by legislation, such as the Greenhouse Gas (GHG) Reporting Program under the Environmental Protection Agency (EPA). Others that have proceeded under the authority of the Clean Air Act and other existing laws include standards on tailpipe emissions of GHG from vehicles. For more information, see CRS Report R41212, EPA Regulation of Greenhouse Gases: Congressional Responses and Options, by James E. McCarthy.

[2] White House, "The President's Climate Action Plan," June 2013, http://www.whitehouse.gov/ sites/default/files/ image/president27sclimateactionplan.pdf. For more discussion, see CRS Report R43120, President Obama's Climate Action Plan, coordinated by Jane A. Leggett.

[3] For example, a program that is the primary financier of a major new technology (e.g., carbon capture and sequestration) generally would require more funds to achieve its intended effects than a program that analyzes, promulgates, and enforces a regulation (depending on the scope and form of the regulation).

[4] OMB's most recent report (Office of Management and Budget. Federal Climate Change Expenditures Report to Congress. Washington, D.C: Executive Office of the President, August 2013) was released August 28, 2013 in response to Title IV, Division E, Section 425 of the Consolidated Appropriations Act of 2012 (P.L. 112-74), and continued under Consolidated and Further Continuing Appropriations Act, 2013 (P.L. 113-6).

[5] OMB's compilation largely eliminates issues with double-counting.

[6] For more detailed discussion of reporting challenges and examples, see archived CRS Report RL33817, Climate Change: Federal Program Funding and Tax Incentives, by Jane A. Leggett, as well as GAO. Climate Change: Improvements Needed to Clarify National

Priorities and Better Align Them with Federal Funding Decisions. Washington DC, May 20, 2011. http://www.gao.gov/products/GAO-11-317.

[7] According to OMB, in Budget System and Concepts (2008), "budget authority" is "the authority provided by law to incur financial obligations that will result in outlays," or spending. It should not be confused with actual expenditures of funds. See http://www.whitehouse.gov/omb/budget/fy2008/pdf/concepts.pdf. Budget authority identified in this memorandum for a given year is "new budget authority"—which this report uses interchangeably with "funding." New budget authority differs from total budgetary resources, or the total budget authority available for a specific purpose in a given year. (In some contexts, the term budget authority may be used to include new BA plus residual BA left unobligated, or unspent, from previous years.)

[8] For more information, see U.S. Office of Management and Budget, Federal Climate Change Expenditures Report to Congress, June 2010, at http://www.whitehouse.gov/sites/default/files/omb/assets/legislative_reports/ FY2011_Climate_Change.pdf.

[9] See GAO. Climate Change: Federal Reports on Climate Change Funding Should Be Clearer and More Complete. Washington DC, August 25, 2005. http://www.gao.gov/ products/GAO-05-461.

[10] The Agency for International Development does not receive direct budget authority, but receives funding from other USGCRP agencies to cooperate in a few projects in low-income countries, for example, to support development of drought early warning systems. The Department of Defense also performs research related to global climate change, but those amounts are not included in the budget request for the USGCRP.

[11] For more information, see CRS Report R41845, The Global Climate Change Initiative (GCCI): Budget Authority and Request, FY2010-FY2014, by Richard K. Lattanzio.

[12] GAO, Improvements Needed to Clarify National Priorities and Better Align Them with Federal Funding Decisions GAO-11-317, May 20, 2011; GAO, Climate Change: Greater Clarity and Consistency Are Needed in Reporting Federal Climate Change Funding, GAO-06-1122T, Washington DC, September 21, 2006; GAO. Climate Change: Federal Reports on Climate Change Funding Should Be Clearer and More Complete. Washington DC, August 25, 2005.

[13] GAO. Climate Change: Improvements Needed to Clarify National Priorities and Better Align Them with Federal Funding Decisions. Washington DC, May 20, 2011. http://www.gao.gov/products/GAO-11-317. Highlights.

[14] For another federal cross-cut budget, see discussion in CRS Report RL34329, Crosscut Budgets in Ecosystem Restoration Initiatives: Examples and Issues for Congress, by Pervaze A. Sheikh and Clinton T. Brass.

In: Federal Climate Change Funding
Editor: Sandrin Toullart
ISBN: 978-1-62948-554-6

Chapter 2

FEDERAL CLIMATE CHANGE EXPENDITURES REPORT TO CONGRESS*

Office of Management and Budget

1. INTRODUCTION

"We can't have an energy strategy for the last century that traps us in the past. We need an energy strategy for the future – an all-of-the-above strategy for the 21st century that develops every source of American-made energy."

—President Barack Obama, March 15, 2012

"We will continue to lead by the power of our example, because that's what the United States of America has always done. I am convinced this is the fight America can, and will, lead in the 21st century. And I'm convinced this is a fight that America must lead. But it will require all of us to do our part. We'll need scientists to design new fuels, and we'll need farmers to grow new fuels. We'll need engineers to devise new technologies, and we'll need businesses to make and sell those technologies. We'll need workers to operate assembly lines that hum with high-tech, zero-carbon components, but we'll also need builders to hammer into place the foundations for a new clean energy era."

—President Barack Obama, June 25, 2013

* This is an edited, reformatted and augmented version of the Office of Management and Budget, dated August 2013.

The following is an accounting of Federal funding for climate change programs and activities, both domestic and international, included in the fiscal year (FY) 2014 President's Budget. This report is provided in response to Title IV, Division E, Section 425, of P.L. 112-74, the Consolidated Appropriations Act of 2012 continued under P.L. 113-6, Consolidated and Further Continuing Appropriations Act, 2013:

> Not later than 120 days after the date on which the President's fiscal year 2013 budget request is submitted to Congress, the President shall submit a comprehensive report to the Committee on Appropriations of the House of Representatives and the Committee on Appropriations of the Senate describing in detail all Federal agency funding, domestic and international, for climate change programs, projects and activities in fiscal year 2011, including an accounting of funding by agency with each agency identifying climate change programs, projects and activities and associated costs by line item as presented in the President's Budget Appendix, and including citations and linkages where practicable to each strategic plan that is driving funding within each climate change program, project and activity listed in the report.

1.1. Background

The U.S. Government's portfolio of climate change programs and cross-cutting initiatives focuses on advancing our understanding of climate change and its impact on our communities; advancing the development and introduction of energy-efficient, renewable, and other low- or non-emitting technologies; improving standards for measuring and registering emissions reductions and supporting preparedness and resilience to climate change impacts. Many elements of the Administration's climate change portfolio are designed to provide incentives for greenhouse gas (GHG) emissions reductions domestically to support community-based preparedness and resilience efforts, to ensure that Federal operations and facilities continue to protect and serve citizens in a changing climate, and to promote international initiatives focused on concrete actions toward reducing greenhouse gas emission and enhance climate preparedness globally. The Obama Administration has set a U.S. GHG emissions reduction target in the range of 17 percent below 2005 levels by 2020 and approximately 83 percent below 2005 levels by 2050.

Climate and Global Change Research and Education.

Through the U.S. Global Change Research Program (USGCRP), U.S. scientists are conducting world-class research on climate and global change. The USGCRP coordinates scientific research across 13 Federal departments and agencies with the mission of "build[ing] a knowledge base that informs human responses to climate and global change through coordinated and integrated Federal programs of research, education, communication, and decision support."[1]

Reducing Emissions through Clean Energy Investments and Standards

The Administration is pursuing a wide range of initiatives that reduce greenhouse gas emissions through clean energy technologies and policies. The Administration has made the largest clean energy investment in American history and these investments have allowed the U.S. to double America's renewable power generation since 2008.

International Leadership

Under President Obama's leadership, the United States has engaged the international community to promote sustainable economic growth and to meet the climate change challenge through a number of important venues including: international climate negotiations in Copenhagen (2009), Cancun (2010), and Durban (2011); the Major Economies Forum, the Clean Energy Ministerial, the Climate and Clean Air Coalition, and the Asia-Pacific Economic Cooperation (APEC) Summit.

Climate Change Adaptation

At the request of President Obama, an interagency Climate Change Adaptation Task Force has crafted recommendations for how Federal agency policies and programs can better prepare the United States to address the risks associated with a changing climate. Federal agencies have released their first-ever climate change adaptation plans to help ensure smart decisions that protect our investments and safeguard the health and security of our communities, economies, natural resources, and infrastructure from the impacts of severe weather, rising sea levels, and other changing climate conditions. The Task Force has also helped develop the National Fish Wildlife and Plants Climate Adaptation Strategy to guide ecosystem adaptation and resiliency efforts.[2]

The budget information presented in this report reflects the Administration's commitment to address climate change while preserving a strong American economy. The President's 2014 Budget proposes over $21.4 billion for climate change activities. This amount is $1.2 billion, or 5 percent, lower than the 2013 enacted level for climate change programs, activities, and related tax policies.

1.2. Report Outline

The President's 2014 Budget supports a wide range of climate change-related research, development, and deployment programs, voluntary partnerships, and international aid efforts. This report presents the expenditures associated with this portfolio of activities in five main categories – science, technology, international assistance, tax provisions, and adaptation efforts associated with natural resource adaptation – as described below:

- **Climate Change Science.** This category encompasses the U.S. Global Change Research Program (USGCRP).
- **Clean Energy Technology.** Clean Energy Technology incorporates a variety of technology research, development, and deployment activities – including voluntary partnerships and grant programs – that support reductions in greenhouse gas emissions and reliance on fossil fuels. This category comprises work on clean energy systems and sources such as geothermal, solar, wind, biomass, nuclear, and emerging sources such as water power. It also includes programs or technologies or practices that help improve energy efficiency or reduce energy consumption, such as building efficiency, more effective transmission or distribution of electricity, and vehicle technologies that improve engine efficiency or fuel economy.
- **International Assistance.** This category describes elements of a "whole of government" approach to mobilize a wide range of resources and make use of bilateral and multilateral assistance tools. The core budget includes resources for a coordinated set of programs designed to ensure an effective balance across the three pillars of the global climate effort: Adaptation, Clean Energy, and Sustainable Landscapes.
- **Energy Tax Provisions.** This category includes tax incentives for investments in certain energy technologies, and energy payments that can be used in lieu of certain tax credits. These incentives promote

deployment of energy efficient or alternative energy technologies, which may help reduce greenhouse gas emissions.

- **Climate Change Adaptation, Preparedness, and Resilience.** There are numerous efforts across the Federal Government for preparing and building resilience to the impacts of climate change on various critical sectors, institutions, and agency mission responsibilities. This concept is also known as "adaptation." Led by the Interagency Climate Change Adaptation Task Force, and using risk management principles, agencies are working to ensure they can continue to perform their missions in the face of climate change. Successful preparedness efforts often involve integrating climate change considerations into existing agency programs, projects, and activities rather than establishing separate and distinct programs. This creates a challenge when attempting to fully account for all adaptation resources. While the Administration continues to develop methodologies to account for a broader suite of adaptation programs across all critical sectors, an interim category, described further in section 6, summarizes certain activities at the Department of the Interior designed to promote preparedness and resilience. The activities at the Department of the Interior reflect interagency efforts to address key adaptation challenges that cut across the jurisdictions and missions of individual Federal agencies, and affect fresh water, oceans and coasts, and fish, wildlife and plants.

The following sections provide further detail in each of these five areas.

2. Climate Change Science

The U.S. Global Change Research Program (USGCRP) was mandated by Congress in the Global Change Research Act of 1990 (P.L. 101-606) to improve understanding of uncertainties in climate science, including the cumulative effects on the environment of human activities and natural processes, develop science-based resources to support policymaking and resource management, and communicate findings broadly among scientific and stakeholder communities. Thirteen departments and agencies participate in the USGCRP. The Office of Science and Technology Policy (OSTP) and the Office of Management and Budget (OMB) work closely with the USGCRP to align the research priorities and funding plans with the Administration's priorities and agency plans. The program recently issued a new strategic plan (see description and link below).

The 2014 Budget proposes $2.7 billion for the USGCRP to support the goals set forth in the program's current strategic plan. These activities can be grouped under the following areas: improve our knowledge of Earth's past and present climate variability and change; improve our understanding of natural and human forces of climate change; improve our capability to model and predict future conditions and impacts; assess the Nation's vulnerability to current and anticipated impacts of climate change; and improve the Nation's ability to respond to climate change by providing climate information and decision support tools that are useful to policymakers and the general public. Reports and general information about the USGCRP are available on the program's website, www.globalchange.gov.

Table 1. Summary of Federal Climate Change Expenditures (budget authority in millions of dollars)

Summary of Climate Expenditures[1]	**FY 2012 Enacted Budget Authority**	**FY 2013 Enacted Budget Authority**	**FY 2013 Enacted Budget Authority**[8]	**FY 2014 Enacted Budget Authority**	**Change in Budget Authority 2013-2014**
US Global Change Research Program (USGCRP)	2,506	2,509	2,463	2,658	+149
Clean Energy Technologies	6,121	6,088	5,783	7,933	+1,845
International Assistance [2,7]	958	851	797	893	+42
Natural Resources Adaptation	88	95	95	110	+15
Energy Tax Provisions That May Reduce Greenhouse Gases [3,4]	5,052	4,999	4,999	5,129	+130
Energy Payments in Lieu of Tax Provisions [5,6]	5,080	8,080	8,080	4,710	-3,370
Adjustments for programs included in multiple categories	*-24*	*-24*	*-22*	*-23*	---
Total [1,7]	**19,781**	**22,598**	**22,195**	**21,408**	**-1,189**

[1] Budget Authority provided in millions of dollars and are current as of June 21, 2013. Discrepancies with other published documents may result from rounding and improved estimates.

[2] International Assistance includes congressionally appropriated assistance by core agencies (i.e. Department of State, Department of Treasury, US Agency for International Development) as well as complementary agencies (e.g., Environmental Protection Agency), but does not include indirect climate assistance nor development finance and export credit agencies.

[3] Tax incentives related to climate change included in this report were projected at about $23.5 billion over five years (2014-2018). These estimates do not reflect the extension of several temporary tax provisions by the American Taxpayer Relief Act of 2012.

[4] Tax expenditures are estimates of the revenue losses due to a tax preference. While not exactly equivalent to budget authority, tax expenditure estimates are included for completeness.

[5] Firms can take an energy payment in lieu of certain tax credits. The payments are considered outlays and are direct substitutes for the energy tax provisions. Estimates have been included in all columns for completeness.

[6] Energy payments in lieu of tax credits included in this report are currently projected at $9.1 billion over five years (2014-2018).

[7] The International Assistance total contains funds that are also counted in the USGCRP and Clean Energy Technology totals. Table total line excludes this double-count.

[8] Current Budget Authority for FY 2013 throughout this document reflects the amount the program has available for the year calculated as the appropriated amount (as reported in the FY 2013 Enacted column) minus the reductions pursuant to the Budget Control Act of 2011 (P.L. 112-25) sequestration order issued on March 1, 2013, and accounting for any known and applicable reprogrammings, transfers, or other related adjustments. Estimates are current as of June 21, 2013 and are subject to change.

2.1. Selected Agency Highlights of the USGCRP in the 2014 Budget

- **Understand and Accurately Project Climate Change and its Impacts.** The U.S. Global Change Research Program (USGCRP) integrates Federal research and solutions for climate and global change. The new strategic plan will guide interagency investments in the Budget, including support for a National Climate Assessment of the current science and impacts of climate change. The Budget provides nearly $2.7 billion for USGCRP programs, an increase of $147 million (6 percent) above the FY 2013 enacted level.

- The Department of Commerce's National Oceanic and Atmospheric Administration (NOAA) is a leading sponsor of oceanic and atmospheric research and is one of the key sponsors of climate science capabilities in the Federal government. The 2014 Budget allocates $371 million for the Department of Commerce's USGCRP efforts, predominantly from NOAA; this represents an increase of $55 million or 17 percent over the FY 2013 enacted level.
- The National Aeronautics and Space Administration's (NASA) budget includes a sustained investment in climate science, with $1.5 billion proposed for FY 2014. NASA's Earth Science program conducts first-of-a-kind demonstration flights of sensors in air and space in an effort to foster scientific understanding of the Earth system and to improve the ability to forecast climate change and natural disasters. The 2014 Budget supports several research satellites in development, an initiative to monitor changes in polar ice sheets, enhancements to climate models, and NASA contributions to the USGCRP's National Climate Assessment. NASA will continue to develop a replacement to the Orbiting Carbon Observatory (OCO).
- The National Science Foundation (NSF) provides funding for academic basic research across the entire spectrum of the sciences, engineering, and the social sciences. NSF USGCRP support totals $326 million in the 2014 Budget.
- The Department of Energy (DOE) conducts research on climate modeling and predictability that also involves advancing climate and earth system models with improved resolution and uncertainty quantification; DOE also supports long-term atmospheric and terrestrial research experiments. The 2014 Budget allocates $220 million coordinated through USGCRP, with a $7 million increase over FY 2013 dedicated to major field experiments at Arctic, tropics, and oceanic sites. DOE also partners with NSF to support the Community Earth System Model.
- The 2014 Budget provides $72 million for USGCRP programs in the Department of the Interior, an increase of $14 million or 24 percent over the 2013 funding level. Interior's lead science agency, the U.S. Geological Survey (USGS), funds several programs in coordination with other USGCRP agencies to

understand the impacts of climate change on natural resources, including the National Climate Change and Wildlife Science Center, which supports a network of Climate Science Centers (CSCs). The CSC supports development of actionable science linked to resource management decisions on climate adaptation.

2.2. Linkages to Strategic Plans

Interagency Strategic Plans

- *USGCRP 2012-2021 Strategic Plan.* This ten-year interagency strategic plan is built around four strategic goals: Advance Science, Inform Decisions, Conduct Sustained Assessments, and Communicate and Educate. In addition to these four goals, the plan emphasizes the importance of national and international partnerships that leverage Federal investments and provide for the widest use of program results. The plan builds on the program's strengths in integrated observations, modeling, and information services for science that serves societal needs. http://downloads.globalchange.gov/strategic-plan/2012/usgcrp-strategic-plan-2012.pdf
- *Our Changing Planet.* Since 1989 the Global Change Research Program has submitted an annual report to Congress summarizing recent achievements, near term plans, and progress in implementing long term goals. *Our Changing Planet* also provides an overview of recent and near-term expenditures and of requested funding. http://library.globalchange.gov/products/annualreports

Individual Agency Strategic Plans. Excerpts from each participating Agency's strategic plans are provided below along with a weblink to each respective strategic plan.

- *Department of Agriculture.* Climate change is a central consideration in USDA's strategic planning. Strategic Goal 2 of USDA's Strategic Plan is titled *Ensure our National forests and private working lands are conserved, restored, and made more resilient to climate change, while enhancing our water resources.* USDA also developed a Climate Change Science Plan which presents an overview of the critical questions facing the Department's agencies as they relate to climate change and offers a framework for assessing priorities to ensure consistency with USDA's role in the USGCRP. The objectives of the Climate Change Science Plan include:

- Restoring and conserving the Nation's forests, farms, ranches, and grasslands.
- Leading efforts to mitigate and adapt to climate change.
- Protecting and enhancing America's water resources.
- Reducing risk from catastrophic wildfire and restore fire to its appropriate place on the landscape.
- Supporting ecological restoration of our Nation's forests and grasslands and providing research to support improved forest management.
- http://www.usda.gov/oce/climate_change/science_plan2010/USDA_CCSPlan_120810. pdf

- *Department of Commerce.* Under its broad goals of generating and communicating new, cutting-edge scientific understanding and promoting economically-sound environmental stewardship and science, the Department of Commerce's Strategic Plan highlights several objectives that will accomplish the following:
  - Advance scientific knowledge and understanding of the Earth's systems, its changing climate, and associated impacts; enhance weather, water, and climate reporting and forecasting; integrate assessments of current and future climate that identify potential impacts; support mitigation and adaptation efforts through sustained, reliable, and timely climate services; and inform the public so that it understands its vulnerabilities to a changing climate and makes informed decisions.
  - http://www.osec.doc.gov/bmi/budget/DOC_Strategic_Plan_022311.pdf
- *Department of Energy.* DOE's Strategic Plan includes Goal 2: *Maintain a vibrant U.S. effort in science and engineering as a cornerstone of our economic prosperity with clear leadership in strategic areas*; these areas include climate science. The Strategic Plan describes DOE's climate science objective to support *"basic and policy-relevant research underpinning a predictive, systems-level understanding of climate."* To achieve this goal, DOE will:
  - Support fundamental scientific research on climate predictability for improved future projections at the regional spatial scale and with time scales extending from sub- decadal to centennial as part

of the U.S. Global Change Research Program and in coordination with the international science community.

- Provide long-term support to major field research facilities, involving a combination of experimental and modeling activities that focus on atmospheric clouds and aerosols, and terrestrial ecosystems; many of the DOE investments leverage decades of field experience involving sophisticated observational and analytical expertise that has been deployed to sites extending from the Artic to the tropics.
- Provide long-term support to the comparison, analysis, and diagnosis of all climate models worldwide, in order to enhance US competitiveness in the science of climate predictability.
- http://energy.gov/sites/prod/files/2011_DOE_Strategic_Plan_.pdf

- *Department of Health and Human Services, National Institutes of Health.* The FY 2012- 2017 Strategic Plan for the National Institute of Environmental Health Sciences has as its Strategic Goal 5: *Identify and respond to emerging environmental threats to human health, on both a local and global scale.* To achieve this goal NIEHS will:
  - Focus on research needs to help inform policy response in public health situations in which lack of knowledge hampers policymaking, e.g., to improve understanding of the health effects that result from exposures related to climate change.
  - http://www.niehs.nih.gov/about/strategicplan/strategicplan2012_508.pdf
- *Department of the Interior.* DOI's FY 2011-2016 Strategic Plan contains a strategy to assess and forecast climate change and its effects with its Strategic Goal 2: *Provide Science for Sustainable Resource Use, Protection, and Adaptive Management* and Mission Area 4: *Provide a Scientific Foundation for Decision Making.* This strategy notes that successful adaptation to climate change will depend on access to a variety of options for effective management responses, and describes USGS efforts to:
  - Develop, implement, and test adaptive strategies, reduce risk, and increase the potential for ecological systems to be self-sustaining, resilient, and adaptable to environmental changes.
  - Implement partner-driven science to improve understanding of past and present land use change, develop relevant climate and land use forecasts, and identify lands, resources, and communities

that are most vulnerable to adverse impacts of change from the local to global scales.
    - http://www.doi.gov/pfm/upload/DOI_StrategicPlan_FY11-16.pdf
- *Department of Transportation.* DOT's FY 2012-2016 Strategic Plan includes the Strategic Goal *Advance Environmentally Sustainable Policies and Investments that Reduce Carbon and Other Harmful Emissions from Transportation Sources.* Included in DOT's strategies to achieve this goal are the following:
    - Work through DOT's virtual Center for Climate Change to coordinate climate-related activities, research, and products with the climate experts throughout the Department.
    - Advance aviation climate research to understand the impacts of high-altitude aircraft emissions.
    - Provide technical assistance and incentives to States and Metropolitan Planning Organizations on strategies that reduce GHG emissions.
    - http://www.dot.gov/sites/dot.dev/files/docs/990_355_DOT_StrategicPlan_508lowres.pdf
- *Environmental Protection Agency.* EPA's FY 2011-2015 Strategic Plan identifies climate change science objectives. Potential impacts of climate change may include increased smog in many regions making it difficult to maintain clean air standards. Climate change may also affect water quality as large volumes of water can overload storm and waste water systems. The Agency's Strategic Plan addresses these challenges in its air and water quality goals.
    - Within EPA's Strategic Goal 1: *Taking action on climate change and improving air quality,* the Strategic Plan identifies an applied research effort to investigate the influence of climate change on clean air, as well as the impacts of emissions from low- carbon fuels in transportation.
    - To achieve EPA's Strategic Goal 2: *Protecting America's water*, EPA will begin to identify actions to respond and adapt to the current and potential impacts of climate change on aquatic resources, including impacts associated with warming temperatures, changes in rainfall amount and intensity, and sea level rise.
    - http://www.epa.gov/planandbudget/strategicplan.html
- *National Aeronautics and Space Administration.* The 2011 NASA Strategic Plan states as its second Strategic Goal: *Expand scientific*

understanding of the Earth and the universe in which we live, and as its Outcome (2.1): *Advance Earth system science to meet the challenges of climate and environmental change.* Within this strategic goal NASA's plan describes several objectives related to climate change science, including:

- Improve understanding of and improve the predictive capability for changes in the ozone layer, climate forcing, and air quality associated with changes in atmospheric composition.
- Enable improved predictive capability for weather and extreme weather events.
- Quantify, understand, and predict changes in Earth's ecosystems and biogeochemical cycles, including the global carbon cycle, land cover, and biodiversity.
- Quantify the key reservoirs and fluxes in the global water cycle and assess water cycle change and water quality.
- Improve understanding of the roles of the ocean, atmosphere, land and ice in the climate system and improve predictive capability for its future evolution.
- Characterize the dynamics of Earth's surface and interior and form the scientific basis for the assessment and mitigation of natural hazards and response to rare and extreme events.
- Enable the broad use of Earth system science observations and results in decision- making activities for societal benefits.
- http://www.nasa.gov/pdf/516579main_NASA2011StrategicPlan.pdf

- *National Science Foundation.* NSF's Strategic Plan FY 2011-2016 contains the strategic goal *Innovate for Society* addressing societal needs through research and education, and highlighting the role that new knowledge and creativity play in economic prosperity and society's general welfare. NSF has set Performance Goal (I-1) to *Make investments that lead to results and resources that are useful to society*, and includes a target to:
    - Support research that underpins long-term solutions to societal challenges such as climate change.
    - http://www.nsf.gov/news/strategicplan/nsfstrategicplan_2011_2016.pdf
- *Smithsonian Institution.* SI's FY 2012-2015 Strategic Plan describes its research-related strategic goal to *Advance and synthesize knowledge that contributes to the survival of at-risk ecosystems and*

species. The Strategic Plan states an objective to understand how certain environmental stressors including climate change affect the survival of species and the functioning of ecosystems, and includes the following strategies:

- Enhance the Smithsonian's platforms for long-term research on biodiversity and ecosystems, particularly the Smithsonian Institution Global Earth Observatories (SIGEO).
- Marshal the Smithsonian's critical mass of biologists and paleontologists, in partnership with experts in other disciplines, to develop understanding of species and ecosystems and find innovative approaches to the complex meta-problems of biodiversity loss, ecosystem degradation, and climate change.
- http://www.si.edu/content/pdf/about/si_strategic_plan_2010-2015.pdf

- *U.S. Agency for International Development.* In the USAID Policy Framework 2011-2015, a core objective is to reduce climate change impacts and promote low emissions growth. This includes the following research effort:
  - Finance up to six regional Earth observation hubs to provide over 30 developing countries with better climate change and forecasting data, enabling them to make better decisions in a wide range of areas likely to be affected by climate change.
  - http://transition.usaid.gov/policy/USAID_PolicyFramework.PDF

Table 2. U.S. Global Change Research Program Details by Agency/Account (Budget authority in millions of dollars)[1]

U.S. Global Change Research Program (USGCRP)[1]	FY 2012 Enacted Budget Authority	FY 2013 Enacted Budget Authority	FY 2013 Current Budget Authority[2]	FY 2014 Proposed Budget Authority	Proposed Change in Budget Authority 2013-2014
Department of Agriculture					
Agricultural Research Service	36	36	38	52	+16
National Institute of Food and Agriculture	50	40	40	43	+3
Economic Research Service	2	2	2	2	----

U.S. Global Change Research Program (USGCRP)[1]	FY 2012 Enacted Budget Authority	FY 2013 Enacted Budget Authority	FY 2013 Current Budget Authority[2]	FY 2014 Proposed Budget Authority	Proposed Change in Budget Authority 2013-2014
Forest Service – Forest and Rangeland Research	26	25	25	28	+3
National Agricultural Statistics Service	1	1	1	1	----
Natural Resources Conservation Services	1	1	1	1	----
Subtotal – USDA[3]	**116**	**104**	**106**	**126**	**+22**
Department of Commerce					
National Oceanic and Atmospheric Administration –Operations, Research, and Facilities	245	247	233	307	+60
National Oceanic and Atmospheric Administration – Procurement, Acquisition, and Construction	69	64	64	59	-5
National Institute of Standards and Technology (NIST)	5	5	5	5	----
Subtotal – DOC[3]	**319**	**316**	**302**	**371**	**+55**
Department of Energy					
Science – Biological & Environmental Research	**211**	**213**	**209**	**220**	**+7**
Department of Health and Human Services					
Centers for Disease Control and Prevention	6	7	7	7	
National Institutes of Health	8	8	8	8	
Subtotal – HHS[3]	**14**	**15**	**14**	**15**	
Department of the Interior					
U.S. Geological Survey – Surveys, Investigations, and Research	**59**	**58**	**55**	**72**	**+14**
Department of Transportation					
Federal Highway Administration – Federal-Aid Highways[4]	0	0	0	0	----
Federal Aviation Administration – Research, Engineering, and Development	1	1	1	1	----
Federal Transit Administration - Research and University Research Centers[5]	0	0	0	0	----
Subtotal – DOT[3]	**1**	**1**	**1**	**1**	----
Environmental Protection Agency					
Science and Technology	**18**	**19**	**17**	**20**	**+1**

Table 2. (Continued)

U.S. Global Change Research Program (USGCRP)[1]	FY 2012 Enacted Budget Authority	FY 2013 Enacted Budget Authority	FY 2013 Current Budget Authority[2]	FY 2014 Proposed Budget Authority	Proposed Change in Budget Authority 2013-2014
National Aeronautics and Space Administration					
Science	**1,427**	**1,447**	**1,435**	**1,499**	**+52**
National Science Foundation					
Research and Related Activities	**333**	**328**	**316**	**326**	**-2**
Smithsonian Institution					
Salaries and Expenses	**8**	**8**	**8**	**8**	----
U.S. Agency for International Development					
Development Assistance- non-add[6]	*11*	*11*	*11*	*14*	*+3*
Department of State					
Other- non-add[7]	*3*	*3*	*3*	*3*	**----**
Total[3]	**2,506**	**2,509**	**2,463**	**2,658**	**+149**

[1] All data supersede numbers released with the 2014 Budget and are current as of June 21, 2013. Budget authority provided in millions of dollars. Any discrepancies are the result of rounding and improved estimates.

[2] Current Budget Authority for FY 2013 throughout this document reflects the amount the program has available for the year calculated as the appropriated amount (as reported in the FY 2013 Enacted column) minus the reductions pursuant to the Budget Control Act of 2011 (P.L. 112-25) sequestration order issued on March 1, 2013, and accounting for any known and applicable reprogrammings, transfers, or other related adjustments. Estimates are current as of June 21, 2013 and are subject to change.

[3] Agency subtotals and table total may not add due to rounding.

[4] The FY 2012 through FY 2014 funding for Federal Highway Administration – Federal Highway Administration – Federal-Aid Highways was less than $500,000.

[5] Federal Transit Administration – Research and University Research Centers is FTA's support for DOT's Center for Climate Change. The FY 2012 through FY 2014 funding amounts for this program are less than $500,000.

[6] USAID funding supports USGCRP and the Climate Change International Assistance effort. In the past, some USAID funding was counted under both categories. These efforts do not add to the USGCRP total.

[7] These efforts do not add to the USGCRP total.

3. CLEAN ENERGY TECHNOLOGIES

Clean Energy Technologies help to reduce, avoid, or sequester greenhouse gas emissions. These programs comprise research, development, and deployment efforts, including a variety of voluntary partnership and grant activities. The activities have the effect of stimulating the development and use of certain energy technologies, including renewable, low-carbon fossil, and nuclear technologies as well as energy efficient technologies, products, and process improvements.

Building on the Administration's progress to make the U.S. the global leader in the clean energy race and protect the environment for generations to come, the 2014 Budget will support American leadership in clean energy. Moving toward a clean energy economy will improve the air we breathe and the water we drink and enhance our energy security by reducing dependence on oil. Clean energy will play a crucial role in slowing global climate change and meeting the President's goals of cutting greenhouse gas emissions in the range of 17 percent below 2005 levels by 2020, and 83 percent by 2050. Just as important, ensuring that the Nation leads the world in the clean energy economy is an economic imperative.

The 2014 Budget proposes approximately $7.9 billion for Clean Energy Technologies. Table 3 provides a breakdown by agency of Clean Energy Technology funding.

Descriptions of some select activities are included below.

3.1. Selected Agency Highlights of Clean Energy Technologies

- **Increased Investment in DOE Climate Change Technology activities.** The Budget proposes $6.2 billion for clean energy technology programs at the Department of Energy, 44 percent more than the 2013 enacted level. The Department's funding supports a wide range of important research, development, and deployment activities on key technologies such as solar, wind, nuclear, and carbon capture and storage. Highlights include:
  - $2.8 billion for the Office of Energy Efficiency and Renewable Energy (EERE) to accelerate research and development, to build on ongoing successes, and to further reduce the costs and increase the use of critical clean energy technologies. Within EERE, the Budget invests $957 million to increase the affordability and

convenience of advanced vehicles and domestic renewable fuels and $615 million in innovative projects to make clean, renewable power, such as solar energy and off-shore wind, more easily integrated into the electric grid and as affordable as electricity from conventional sources, without subsidies. It also more than doubles funding to $885 million for energy efficiency and advanced manufacturing activities to help reduce energy use and costs in commercial and residential buildings, in the industrial and business sectors, and in Federal buildings and fleets.

- $379 million for the Advanced Research Projects Agency - Energy (ARPA-E) to support transformational research in clean energy in areas such as solar energy, energy storage, carbon capture and storage, and advanced biofuels.

3.2. Linkages to Strategic Plans

Interagency Strategic Plans

- *Blueprint for a Secure Energy Future.* In March 2011, the Obama Administration released the *Blueprint for a Secure Energy Future* which outlines the comprehensive national energy policy pursued by the Administration. The *Blueprint* describes strategies across the Federal Government aimed to develop and secure America's energy supplies, provide consumers with choices to reduce costs and save energy and innovate to a clean energy future. http://www.whitehouse.gov/sites/default/files/blueprint_secure_energy_future.pdf
- *Secure Energy Future: Progress Report.* In March 2012 accomplishments and achievements that underscore the Administration's commitment to promoting clean energy technologies were described in the *Secure Energy Future: One Year Progress Report.* This report highlights efforts to increase energy independence, set historic fuel economy standards, improve energy efficiency, expand renewable fuel generation, develop advanced alternative fuels and support cutting-edge research. http://www.whitehouse.gov/sites/default/files/email-files/the_blueprint_for_a_secure_energy_future_oneyear_progress_report.pdf

Table 3. Clean Energy Technologies
Details by Agency/Account (Budget authority in millions of dollars)[1]

Clean Energy Technologies[1]	FY 2012 Enacted Budget Authority	FY 2013 Enacted Budget Authority	FY 2013 Current Budget Authority[10]	FY 2014 Proposed Budget Authority	Proposed Change in Budget Authority 2013-2014
Department of Agriculture					
Natural Resources Conservation Service – Conservation Operations	6	6	0	4	-2
Agricultural Research Service – Salaries and Expenses	33	32	32	39	+7
National Institute of Food and Agriculture - Research and Education Activities	31	57	56	51	-5
Forest Service – Commercialization/ Renewable Energy	26	23	23	28	+5
Rural Business Cooperative Service – Value Added Producer Grants (Cooperative Development Grants)	1	2	1	1	-1
Rural Business Cooperative Service – Rural Energy Program Account (Rural Energy for America Sec. 9007)	3	3	3	20	+16
Rural Business Cooperative Service – Guaranteed Business and Industry Loans	4	6	5	6	----
Rural Business Cooperative Service – Rural Economic Development Loans[3]	0	0	0	0	----
Economic Research Service[4]	2	2	2	2	----
Office of the ChiefEconomist -Salaries and Expenses[5]	4	3	3	4	+1
Rural Utilities Service -High Cost Energy Grants[6]	4	4	4	0	-4
2008 Farm Bill, Mandatory Funding					
Rural Business Cooperative Service – Rural Energy Program Account (Rural Energy for America Sec. 9007)	22	0	0	70	+70
National Institute of Food and Agriculture – Biomass Research and Development (Sec. 9008)	40	0	0	26	+26

Table 3. (Continued)

Clean Energy Technologies[1]	**FY 2012 Enacted Budget Authority**	**FY 2013 Enacted Budget Authority**	**FY 2013 Current Budget Authority**[10]	**FY 2014 Proposed Budget Authority**	**Proposed Change in Budget Authority 2013-2014**
Farm Service Agency – Biomass Crop Assistance Program	17	0	0	0	----
Farm Service Agency – Commodity Credit Corporation	0	170	161	0	-170
Natural Resources Conservation Service – Farm Security and Rural Investment Programs	16	14	14	14	----
Rural Business Cooperative Service – Energy Assistance Payments (formerly titled Bioenergy Program for Advanced Biofuels (Sec. 9005))	65	0	0	0	----
Subtotal – USDA discretionary funding	*116*	*138*	*130*	*155*	*+18*
Subtotal – USDA mandatory funding	*160*	*184*	*175*	*110*	*-74*
Subtotal – USDA[7]	**275**	**322**	**305**	**265**	**-57**
Department of Commerce					
National Institute of Standards and Technology (NIST) – Scientific and Technological Research and Services	40	40	40	40	----
National Oceanic and Atmospheric Administration Operations, Research and Facilities	0	0	0	3	+3
Subtotal – Commerce[7]	**40**	**40**	**40**	**43**	**+3**
Department of Defense					
Research, Development, Test and Evaluation, Army	32	29	29	32	+2
Research, Development, Test and Evaluation, Navy	231	186	176	226	+40
Research, Development, Test and Evaluation, Air Force	118	203	190	153	-50
Research, Development, Test and Evaluation, Defense Wide	101	46	42	46	----

Clean Energy Technologies[1]	FY 2012 Enacted Budget Authority	FY 2013 Enacted Budget Authority	FY 2013 Current Budget Authority[10]	FY 2014 Proposed Budget Authority	Proposed Change in Budget Authority 2013-2014
Subtotal – DOD[7]	**481**	**465**	**437**	**457**	**-8**
Department of Energy					
Energy Efficiency and Renewable Energy	1,819	1,810	1,719	2,788	+978
Electricity Delivery and Energy Reliability	133	133	126	153	+20
Nuclear Energy	772	765	723	733	-32
Fossil Energy R&D – Carbon Capture and Storage (CCS) and Power Systems	472	446	425	375	-71
Science – Fusion, Sequestration, and Hydrogen	902	924	883	1,067	+143
Energy Transformation Acceleration Fund -Advance Research Projects Agency-Energy(ARPA-E)	275	264	251	379	+114
Bonneville Power Administration Fund[9]	15	17	17	17	
Race to the Top for Energy Efficiency and Grid Modernization	0	0	0	200	+200
HomeStar	0	0	0	300	+300
Energy Security Trust	0	0	0	200	+200
Subtotal – DOE[7]	**4,388**	**4,359**	**4,144**	**6,212**	**+1,853**
Department of Transportation					
National Highway Traffic Safety Administration- Operations and Research	10	10	8	11	+1
Research and Innovative Technology Administration – Research and Development	1	1	1	1	----
Federal Aviation Administration - Research, Engineering and Development	21	17	20	18	+1
Federal Aviation Administration -Facilities and Equipment	7	5	4	5	+1
Federal Transit Administration - Research and University Research Centers and Formula and Bus Grants	52	23	22	15	-8
Federal Railroad Association - Railroad Research and Development	1	2	1	3	+1

Table 3. (Continued)

Clean Energy Technologies[1]	FY 2012 Enacted Budget Authority	FY 2013 Enacted Budget Authority	FY 2013 Current Budget Authority[10]	FY 2014 Proposed Budget Authority	Proposed Change in Budget Authority 2013-2014
Subtotal – DOT[7]	**91**	**57**	**56**	**52**	**-5**
Environmental Protection Agency					
Environmental Programs and Management	99	99	95	106	+7
Science and Technology	18	17	16	10	-7
Subtotal – EPA[7]	**117**	**116**	**111**	**115**	
National Aeronautics and Space Administration					
Aeronautics	259	262	255	284	+22
Exploration	9	7	6	9	+1
Space Technology	28	15	15	28	+14
Subtotal – NASA[7]	**296**	**284**	**276**	**321**	**+37**
National Science Foundation					
Research and Related Activities	**341**	**352**	**346**	**372**	**+20**
Nuclear Regulatory Commission					
Salaries and Expenses[8]	**83**	**82**	**57**	**86**	**+4**
Tennessee Valley Authority					
Tennessee Valley Authority Fund[9]	**9**	**11**	**11**	**10**	**-1**
Total[7]	**6,121**	**6,088**	**5,783**	**7,933**	**+1,845**

[1] All data supersede numbers released with the 2014 President's Budget and are current as of June 21, 2013. Budget authority provided in millions of dollars. Discrepancies may result from rounding and improved estimates.

[3] Funding for the Rural Business Cooperative Service - Rural Economic Development Loans was less than $500,000 in FY 2012 and FY 2013.

[4] USDA's Economic Research Service has been included in the FCCER, this funding is used to conduct research on the economics of renewable energy.

[5] Office of the Chief Economist includes USDA's Climate Change Program Office and The Office of Energy Policy and New Uses (OEPNU) Research and Development

[6] The Rural Utilities Service - High Cost Energy Grants program has activities to support the creation and use of renewable energy and energy efficiencies.

[7] Agency subtotals and table total may not sum due to rounding.

[8] Nuclear Regulatory Commission funding has been included in the FCCER and reflects funding for nuclear energy research.

[9] Tennessee Valley Authority funding has been added to the FCCER and reflects funding for small modular nuclear reactors research as well as R&D relating to the deployment of nuclear technologies, reduction of greenhouse gas emissions, renewable generation, post- combustion carbon dioxide capture technologies, air quality, energy efficiency and demand response.

[10] Current Budget Authority for FY 2013 throughout this document reflects the amount the program has available for the year calculated as the appropriated amount (as reported in the FY 2013 Enacted column) minus the reductions pursuant to the Budget Control Act of 2011 (P.L. 112-25) sequestration order issued on March 1, 2013, and accounting for any known and applicable reprogrammings, transfers, or other related adjustments. Estimates are current as of June 21, 2013 and are subject to change.

Individual Agency Strategic Plans. Excerpts from each participating Agency's strategic plans are provided below along with a weblink to each respective strategic plan.

- *Department of Agriculture.* USDA's Strategic Plan for 2010-2015 includes the objective *Lead efforts to mitigate and adapt to climate change* (Strategic Goal 2, Objective 2.2). The plan also includes an objective, *Enhance rural prosperity,* by facilitating sustainable renewable energy development, promoting energy efficiency, and curbing the effects of climate change. (Strategic Goal 1, Objective 1.1). These objectives includes numerous efforts and strategies including:
    - Providing assistance to farmers, ranchers, and forest landowners to implement conservation, nutrient management, and animal management practices that reduce emissions and sequester carbon.
    - Planting and maintaining vegetative cover on marginal farmland and land that has been impacted by fire.
    - Providing assistance in the form of payments, grants, loans and loan guarantees for clean and renewable energy projects and energy efficiency improvements.
    - http://www.ocfo.usda.gov/usdasp/sp2010/sp2010.pdf
- *Department of Commerce.* The Commerce FY 2011-2016 Strategic Plan includes Objective 5: *Provide the measurement tools and standards to strengthen manufacturing, enabling innovation, and*

enhancing efficiency, Objective 6: *Promote and support the advancement of green and blue technologies and industries*, and Objective 16: *Support climate adaptation and mitigation.* Contributing to this objective are the following:

- Focusing on programs at National Institute of Standards and Technology (NIST) that will develop the measurements, standards, and common framework that are required to promote sustainable operations and improve energy efficiency in both the construction and manufacturing sectors.
- Commerce: (http://www.osec.doc.gov/bmi/budget/DOC_Strategic_Plan_022311.pdf)
- NIST: (http://www.nist.gov/director/upload/nist-master-3-year-plan-fy2012-fy2014.pdf)

- *Department of Defense.* As one of the Government's largest consumers of energy, the Department of Defense is committed to supporting the Administration's efforts in Clean Energy. DOD's Operational Energy Strategy outlines three principles for a stronger force: 1) *Reduce the demand for energy in military operations;* 2) *Expand and secure the supply of energy to military operations;* 3) *Build energy security into the future force.* As part of the effort to act on these principles, DOD's Operational Energy Strategy describes goals to:
  - Reduce energy demand, the most immediate operational energy priority for the Department, by investing in new technologies and equipment.
  - Expand supply options, both for near-term tactical benefits and long-term operational energy security.
  - Take energy into account in order to make more informed decisions about the choices and tradeoffs in equipping and employing forces.
  - http://energy.defense.gov/OES_report_to_congress.pdf
- *Department of Energy.* DOE's 2011 Strategic Plan lays a framework for utilizing the DOE's capabilities to drive solutions across energy, environmental, climate, and security challenges. It demonstrates strong linkages between clean energy and progress on environmental issues, such as climate change. Shifting to a clean energy economy directly supports the Administration's climate change objective to reduce energy-related greenhouse gas emissions. In particular, Goal 1 of the Strategy: *Catalyze the timely, material, and efficient*

transformation of the nation's energy system and secure U.S. leadership in clean energy technologies focuses on activities that support transforming the nation's energy system and building a sustainable and competitive clean energy economy. Targeted outcomes that support this goal include:

- DOE and the U.S. Department of Housing and Urban Development working together to enable the cost-effective energy retrofits of a total of 1.1 million housing.
- Double renewable energy generation from wind, solar, and geothermal energy sources.
- Encourage industry to translate our R&D outputs to market through new contractual vehicles that lower transaction costs and address commercialization barriers.
- http://energy.gov/sites/prod/files/2011_DOE_Strategic_Plan_.pdf

- *Department of Transportation.* DOT's FY 2012-2016 Strategic Plan includes a goal to *Advance environmentally sustainable policies and investments that reduce carbon and other harmful emissions from transportation sources.* Contributing to this goal are strategies to:
  - Reduce carbon emissions, improve energy efficiency, and reduce dependence on oil, including establishment of fuel economy standards for cars and trucks and research into alternative aircraft fuels.
  - Reduce transportation-related air, water and noise pollution and impacts on ecosystems, including expanding opportunities for shifting freight from less fuel-efficient modes to more fuel-efficient modes.
  - Increase the use of environmentally sustainable practices in the transportation sector, including more environmentally sound construction and operational practices.
  - Reduce pollution from DOT owned or controlled transportation services and facilities, including implementing net-zero-energy building requirements for all new buildings entering the design process in 2020 and thereafter.
  - Promote the deployment of technologies—such as hydrogen fuel cell and diesel-electric hybrid buses—that reduce the energy consumption and greenhouse gas emissions of transit systems.
  - http://www.dot.gov/sites/dot.dev/files/docs/990_355_DOT_StrategicPlan_508lowres.pdf

- *Environmental Protection Agency.* The first goal in EPA's FY 2011-2015 Strategic Plan, *Taking action on climate change and improving air quality,* includes efforts to:
  - Develop a national system for reporting GHG emissions.
  - Issue standards to reduce emissions from cars and trucks and non-road sources.
  - Implement permitting requirements and voluntary programs to promote energy efficiency and encourage design and construction of more efficient processes.
  - http://www.epa.gov/planandbudget/strategicplan.html
- *National Aeronautics and Space Administration.* The 2011 NASA Strategic Plan states as its Strategic Goal 3: *Create the innovative new space technologies for our exploration, science, and economic future* and Strategic Goal 4: *Advance aeronautics research for societal benefit.* In pursuing these goals NASA is conducting projects that support clean energy technologies as described in several strategic objectives, including:
  - Develop innovative solutions and technologies to meet future capacity and mobility requirements of the Next Generation Air Transportation System (NextGen).
  - Develop tools, technologies, and knowledge that enable significantly improved performance and new capabilities for future air vehicles.
  - Create a pipeline of new innovative concepts and technologies with early-stage Technology Readiness Levels (TRL) for future NASA missions and national needs.
  - Develop advanced technologies to improve the overall safety of the future air transportation system.
  - http://www.nasa.gov/pdf/516579main_NASA2011StrategicPlan.pdf
- *National Science Foundation.* NSF's Strategic Plan for 2011-2016 includes a performance goal to *Make investments that lead to results and resources that are useful to society* (Performance Goal I-1). NSF investments underpin long-term solutions to societal challenges such as economic development, climate change, clean energy, and cyber-security:
  - Near-term actions include expanding partnerships and collaborations with industry or Government agencies in

identifying areas of critical national need and piloting models for investing in priority areas having societal impact.

- Mid-term actions include issuing solicitations and Dear Colleague Letters in areas of critical national need.
- Long-term actions include conducting an impact assessment of the portfolio investments in areas of national need.
- http://www.nsf.gov/news/strategicplan/nsfstrategicplan_2011_2016.pdf

- *Nuclear Regulatory Commission.* Nuclear is considered a clean energy source, and research at NRC to support its regulatory requirements helps maintain nuclear as part of the clean energy mix going forward. Per the NRC's Strategic Plan, the agency's mission is to protect public health and safety, promote the common defense and security, and protect the environment. NRC's strategic goal on safety: *Ensure adequate protection of public health and safety and the environment* supports clean energy technology by:
  - Implementing focused research programs to anticipate and support resolution of safety issues and address new technologies and conduct research programs to identify and support resolution of longstanding and emergent safety issues.
  - http://www.nrc.gov/reading-rm/doc-collections/nuregs/staff/sr1614/v5/sr1614v5.pdf
- *Tennessee Valley Authority.* In August 2010, TVA adopted a renewed vision to be one of the nation's leading providers of low-cost and cleaner energy by 2020. TVA's Strategic Plan supports a shift to a cleaner, more efficient and more diverse generating portfolio providing direction to projects, partnerships, and research and development related to:
  - Idling or retiring aging coal units.
  - Increasing generation from nuclear and renewable resources.
  - Promoting energy efficiency and demand response.
  - Exploring and embracing effective new technologies.
  - http://www.tva.com/abouttva/pdf/TVA4-33149_strategic_plan.pdf

4. International Assistance

The Administration has taken a whole-of-government approach to pursue four broad international climate change financing objectives within its international assistance programs: 1) Demonstrate continued U.S. leadership in forging a global solution to the climate challenge; 2) Help developing countries focus their climate investments strategically over the coming years; 3) Create robust means of measuring, monitoring, and verifying domestic emissions in developing countries; and 4) Reduce vulnerability to climate change. Coordinating and integrating activities from across the U.S. government promotes complementarities that enhance the value of U.S. climate-related financing and increased the likelihood of successfully realizing these four objectives.

Although the Administration's efforts to address climate change are diverse, its bilateral and multilateral international climate change financing is focused on three policy pillars: adaptation, clean energy, and sustainable landscapes. These three policy pillars invest in significant emissions reduction strategies as well as activities that help communities adapt to a changing climate. Key results and indicators to measure progress have been identified for activities in these policy pillars and can be mapped to the Administration's four broad objectives. These activities will strengthen our relationships with other nations, help mitigate the security risk that climate change poses as a threat multiplier in the developing world, support our efforts for a comprehensive, multilateral approach to climate change that involves meaningful actions by all major economies, and create economic opportunities for manufacturers of clean energy technologies.

In its FY 2014 Budget, the Administration is seeking $837 million for core international efforts to combat global climate change, which represents a 5 percent increase from the FY 2013 enacted level. These efforts, conducted by the U.S. Agency for International Development, the U.S. Department of State, and the U.S. Department of the Treasury, will help the most vulnerable countries respond to the growing impacts of climate change, and help forge a global solution to the climate crisis.

The core activities are complemented by an estimated $56 million that is being sought for programs conducted by a range of additional U.S. agencies that address climate change internationally.

In addition to the funding summarized in Table 4, USAID, State, and Treasury will be implementing other programs, such as food security programs or biodiversity programs, in ways that will make a significant contribution to the fight against climate change. Programs focused primarily on non- climate change goals may promote "climate-proofed" development or use adjusted techniques to significantly reduce emissions while promoting other development goals and thereby deliver climate change mitigation and adaptation co-benefits. The Administration estimates the FY 2014 budget authority for these programs to be $216 million.[3] Furthermore, the Administration is enhancing U.S. efforts to address global climate change and promote clean energy technologies in important ways beyond those programs with direct appropriations. Through direct loans, loan guarantees, insurance, and working capital guarantees, U.S. development finance and export credit agencies are increasingly mobilizing investments in clean energy technologies around the world.[4] These U.S. Government financial products will help American firms, financial institutions, and investors, with their foreign partners, address climate change in developing countries, offering global benefits.

Together, these activities will substantially contribute to the international community's renewed efforts to address climate change, including through the implementation of the Copenhagen Accord, and make clear the Administration's commitment to international leadership in the necessary transition to a low emission economy.

4.1. Agency Highlights Regarding International Climate Change Assistance

U.S. Agency for International Development (USAID)

USAID is the lead contributor to bilateral assistance, with a focus on capacity building, civil society building, governance programming, and creating the legal and regulatory environments needed to address climate change. USAID will leverage its significant technical expertise to provide leadership in development and implementation of low-carbon strategies, creating policy frameworks for market-based approaches to emissions reduction and energy sector reform, promoting sustainable management of agriculture lands and forests, and mainstreaming adaptation into development activities in countries most at risk. USAID has long-standing relationships with host country governments that will enable it to develop shared priorities and implementation plans. USAID's engagement and expertise in agriculture, biodiversity, health, and other critical climate sensitive sectors provide an opportunity to implement innovative cross-sector climate change programs.

Finally, USAID bilateral programs can work in key political and governance areas that multilateral agencies cannot.

Department of State

State takes the lead on diplomatic efforts and deploys financial resources in support of key multilateral and bilateral priorities. State's comparative advantage is promoting effective international solutions, advanced technology strategies, and innovative market approaches through international processes and U.S.-led diplomatic partnerships and initiatives.

Department of the Treasury

The Treasury Department is the primary agency through which the U.S. Government provides contributions through multilateral delivery channels, including the Climate Investment Funds and the Global Environment Facility. Multilateral assistance promotes institutional structures governed jointly by developed and developing countries, which are needed for a coordinated, global response to climate change. Multilateral institutions complement bilateral assistance by leveraging contributions from other donors, making capital investments in infrastructure, providing a range of tailored financial products, and working across a larger number of countries.

The FY 2014 Budget requests $216 million for the Clean Technology Fund (CTF), which aims to close the price gap in developing countries between dirtier conventional technologies and commercially available cleaner alternatives in the power sector, the transportation sector, and in energy efficiency. The CTF focus is on transforming energy use on a sector scale in the "larger emitter" developing countries.

In addition, the FY 2014 Budget requests $68 million for the Strategic Climate Fund in three programs: the Pilot Program for Climate Resilience, the Forest Investment Program, the program for Scaling-up Renewable Energy in Low Income Countries (SREP). The Pilot Program for Climate Resilience will help finance comprehensive efforts to improve the technical capacity of countries to plan for and finance climate adaptation efforts. The Forest Investment Program will support activities informed by national plans to reduce deforestation and will focus on transitioning a small number of developing countries to participate in carbon financing for forest preservation. SREP aims to demonstrate how to put the poorest countries on a pathway that uses renewable energy to expand energy access and stimulate economic growth.

Department of Commerce

The Department of Commerce manages the Renewable Energy and Energy Efficiency Export Initiative (RE4I) through its leadership of the TPCC Working Group on Renewable Energy and Energy Efficiency. The RE4I is a key, sector-specific initiative designed to meet the specific needs of the clean energy sector, while also advancing the President's goal of doubling exports. It includes contributions from eight U.S. Government agencies and is meant to facilitate the deployment of renewable energy and energy efficiency (RE&EE) technologies; better link buyers and sellers of RE&EE products and services; open markets for U.S.-made RE&EE technologies; improve U.S. Government financing for RE&EE exporters; and enhance two-way communication between the U.S. Government and the RE&EE industry. Under the auspices of the RE4I, Commerce has facilitated several improvements to the U.S. Government's trade promotion process in the RE&EE sector, including the development of RE&EE trade policy missions to support existing trade promotion activities by helping to create new markets for U.S. RE&EE companies in countries with nascent policy framework and regulatory systems. This has resulted in highly successful missions to Mexico, Japan, Chile, and Saudi Arabia. Commerce has also developed a first-of-its-kind Renewable Energy Top Prospects Study to help the interagency direct trade promotion activities toward those markets most likely to support U.S. exports.

Complementary Agencies

In addition to the core international assistance activities, a number of additional agencies provide technical and in some cases direct support for international efforts to address climate change. Two international agencies, the Millennium Challenge Corporation (MCC) and the U.S. Trade and Development Agency (USTDA), work directly with international partners on projects that may have climate change benefits.

MCC works with its foreign government partners on programs that reduce poverty through sustainable economic growth. In undertaking its poverty reduction programs, MCC seeks to integrate climate change considerations, such as adaptation and reduced emissions, where appropriate. For example, MCC clean energy capital investments may support economic development priorities with ancillary benefits in emissions reductions. MCC agriculture and agricultural infrastructure programs, such as irrigation, may integrate more sustainable use of water resources in those areas at risk of increasing water scarcity in a changing climate.

Table 4. International Climate Change Assistance Details by Agency/Account (Budget authority in millions of dollars)[1]

International Assistance[1,2]	FY 2012 Enacted Budget Authority	FY 2013 Enacted Budget Authority	FY 2013 Current Budget Authority[20]	FY 2014 Proposed Budget Authority	Proposed Change in Budget Authority 2013-2014
Core Agencies[3]					
Department of State					
Diplomatic and Consular Affairs[4]	0	0	0	0	---
Economic Support Fund	96	95	91	94	-1
International Organizations and Programs	37	37	35	39	+2
Subtotal – State[5]	**133**	**132**	**126**	**133**	**+1**
Department of the Treasury[6]					
Debt Restructuring – Tropical Forestry Conservation	12	12	11	0	-12
Global Environment Facility[7, 18]	60	65	63	72	+7
Clean Technology Fund[18]	230	185	175	216	+31
Strategic Climate Fund[8, 18]	75	50	47	68	+18
Subtotal – Treasury[5]	**377**	**311**	**296**	**356**	**+44**
U.S. Agency for International Development[4]					
Assistance for Europe, Eurasia, and Central Asia[9]	15	0	0	0	---
Development Assistance[19]	322	322	308	317	-5
Economic Support Fund	12	28	27	32	+4
International Disaster Assistance	0	0	0	0	---
Subtotal – USAID[5]	**348**	**334**	**335**	**349**	**-1**
Subtotal-Core Agencies[5]	*858*	*792*	*757*	*837*	*+45*
Complementary Agencies[10]					
US Department of Energy[11]					
Energy Efficiency and Renewable Energy	9	9	9	9	---
Fossil Energy R&D –Carbon Capture and Storage (CCS)	3	3	3	3	---
Science	1	1	1	1	---
Subtotal- DOE[5]	**13**	**13**	**13**	**13**	---
Environmental Protection Agency					
Environmental Programs and Management[12]	**18**	**18**	**16**	**19**	---
National Science Foundation					
Research and Related Activities[14]	**6**	**6**	**6**	**3**	**-3**

International Assistance[1,2]	FY 2012 Enacted Budget Authority	FY 2013 Enacted Budget Authority	FY 2013 Current Budget Authority[20]	FY 2014 Proposed Budget Authority	Proposed Change in Budget Authority 2013-2014
Department of Agriculture					
Forest Service-Forest and Rangeland Research[15]	**3**	**3**	**3**	**1**	**-2**
National Aeronautics and Space Administration					
Science[16]	**3**	**3**	**3**	**3**	**---**
Millennium Challenge Corporation					
Millennium Challenge Corporation[17]	**41**	**0**	**0**	**0**	**---**
US Trade and Development Agency					
Trade and Development Agency[21]	**16**	**16**	**0**	**18**	**+1**
Subtotal-Complementary Agencies[5]	*100*	*59*	*40*	*56*	*-3*
Total[5]	**958**	**851**	**797**	**893**	**+42**

[1] This table shows core climate assistance from programs with climate as a primary objective. In addition, indirect climate assistance is provided through programs in other development sectors such as agriculture, water, and health, that do not necessarily have a primary climate objective but nevertheless may provide climate benefits. Those activities have been captured in the U.S. Fast Start Climate Finance Report.

[2] All data supersede numbers released with the 2014 President's Budget and are current as of June 21, 2013. Budget authority provided in millions of dollars. Discrepancies may result from rounding and improved estimates.

[3] Core agencies for the purposes of the Federal Climate Change Expenditure Report are made of the primary climate assistance activities of the Department of State, Department of the Treasury, and US Agency for International Development (USAID). The Federal Climate Change Expenditures Report contained only these core agencies in previous years.

[4] Diplomatic and Consular Affairs continues to support international climate change activities, but because it is not a foreign assistance account, it has been excluded from the international assistance crosscut, beginning in FY 2011.

[5] Agency subtotals and table total may not add due to rounding.

[6] The FY 2012 totals for Treasury climate programming includes a $100 million transfer from the Department of State.

[7] Only 50% of GEF funds are allocated to programs related to climate change and shown here. The full amounts for GEF over respective columns are 2012—120, 2013 enacted—129, 2013 current—125, and 2014 request—144.

[8] The SCF is the second of the multi-donor Climate Investment Funds. It supports three targeted programs: the Pilot Program for Climate Resilience, the Forest

Investment Program, and the Program for Scaling-Up Renewable Energy in Low-Income Countries.

[9] In the 2009 Omnibus appropriation, Congress combined Assistance for Eastern Europe and the Baltic States with Assistance for the Independent States of the Former Soviet Union, making a new account called Assistance for Europe, Eurasia, and Central Asia.

[10] The category of Complementary Agencies was first included in the Federal Climate Change Expenditures Report that followed the FY 2011 President's Budget as a means to account for technical and in some cases direct support for international efforts to address climate change.

[11] DOE funding provides global outreach on advanced clean coal technology and CCS for climate change mitigation and energy security in multilateral forums.

[12] EPA activities include Methane to Markets, International Capacity Building, and contribution to the Multilateral Fund to support the Montreal Protocol on Substances that Deplete the Ozone Layer.

[13] ITA funding represents activities under the Asia Pacific Partnership to promote the development and deployment of cleaner and more efficient energy technologies.

[14] NSF funding is for Basic Research to Enable Agriculture Development (BREAD) through the Directorate for Biological Science.

[15] Forest Service activities include assistance to developing countries to establish and maintain sustainable landscape management.

[16] NASA activities include funding for the SERVIR initiative which consists of two web-based regional monitoring networks to provide environmental (land, sea, atmosphere, biota) information and projections to decision makers in Central America/Caribbean and East Africa.

[17] MCC anticipates applying FY 2013 and FY 2014 funds to support compacts with Ghana, Benin, El Salvador, Morocco, Niger, Tanzania, Liberia, and Sierra Leone, which may include funding to support climate change objectives. Because funds will not be committed until signing of a compact and the projects within each compact are still being developed, MCC cannot yet report FY 2013 and FY 2014 funding that will support climate change objectives.

[18] FY 2012 Enacted includes a $100 million ESF transfer from State to Treasury for the Clean Technology Fund ($45 million), Strategic Climate Fund ($25 million), and the Global Environment Facility ($30 million).

[19] FY 2013 ESF funding includes both ESF Base and ESF OCO funds.

[20] Current Budget Authority for FY 2013 throughout this document reflects the amount the program has available for the year calculated as the appropriated amount (as reported in the FY 2013 Enacted column) minus the reductions pursuant to the Budget Control Act of 2011 (P.L. 112-25) sequestration order issued on March 1, 2013, and accounting for any known and applicable reprogrammings, transfers, or other related adjustments. Estimates are current as of June 21, 2013 and are subject to change.

[21] USTDA provides funding for various forms of investment analysis and technical assistance to promote investment opportunities for U.S. companies in developing countries. USTDA has expanded its clean energy project portfolio dramatically over the last few years.

USTDA has a number of programs that combine support for U.S. exports with a focus on emissions reductions abroad. USTDA provides technical assistance to developing countries on clean energy technologies that U.S. firms provide, organizes visits for foreign entities seeking business opportunities with U.S. firms in the renewable energy sector, and funds studies on future clean energy infrastructure investments.

A number of domestic agencies with significant technical expertise complement the core international assistance activities on climate change through a variety of functions. The Department of Energy and the Environmental Protection Agency provide technical assistance on clean energy investments and environmental regulations undertaken by foreign governments; National Aeronautics and Space Administration and the National Science Foundation provide research and science assistance to the core international assistance agencies that directly supports climate change efforts; and the Forest Service works with USAID on a number of forestry programs that reduce emissions through carbon sequestration.

4.2. Linkages to Strategic Plans

Interagency Strategic Plans and Planning Documents

- *Meeting the Fast Start Commitment – US Climate Finance in Fiscal Year 2012.* The Fast Start document released in November 2012 describes the $7.5 billion provided during the three-year fast start finance period from 2010-2012. The three-year fast start finance total consists of more than $4.7 billion of Congressionally-appropriated assistance and more than $2.7 billion from U.S. development finance and export credit agencies. http://www.state.gov/documents/organization/201130.pdf
- *Fifth Climate Action Report to the UN Framework Convention on Climate Change.* The *U.S Climate Action Report 2010* sets out the major actions the U.S. government is taking at the federal level, highlights examples of state and local actions, and outlines U.S.

efforts to assist other countries' efforts to address climate change. http://www.state.gov/documents/organization/140636.pdf

- *Fact Sheet: U.S. Global Development Policy and Global Climate Change Initiative.* In September 2010, the President signed a Presidential Policy Directive on Global Development, which provides clear policy guidance to all U.S. Government agencies and enumerates the core objectives, operational model, and the modern architecture needed to implement the policy. http://www.whitehouse.gov/sites/default/files/Climate_Fact_Sheet.pdf

Individual Agency Strategic Plans and Planning Documents

- *USAID.* The *Climate Change and Development Strategy/2012-2016*, released in January 2012, describes USAID's efforts to enable countries to accelerate their transition to climate resilient low emission sustainable economic development. To accomplish this, USAID will pursue three strategic objectives:
  - Accelerate the transition to low-emission development through investments in clean energy and sustainable landscapes;
  - Increase resilience of people, places, and livelihoods through investments in adaptation;
  - Strengthen development outcomes by integrating climate change in Agency programming.
  - http://transition.usaid.gov/our_work/policy_planning_and_learning/documents/GC CS.pdf
- *State and USAID.* The *Quadrennial Diplomacy and Development Review,* released in 2010, describes the whole-of-government approach used by the Global Climate Change Initiative (GCCI). Key GCCI objectives include:
  - Laying the foundation for low-carbon growth by supporting partner country efforts to advance economic growth while reducing emissions;
  - Accelerating the clean energy revolution through multilateral and bilateral mechanisms and promoting development and deployment of clean energy technologies;
  - Reducing emissions from agricultural and other land use and conserving forests through contributions to Reducing Emissions from Deforestation and Forest Degradation (REDD+).
  - http://www.state.gov/s/dmr/qddr/index.htm

5. Energy Tax Provisions That May Reduce Greenhouse Gases

This report includes existing energy tax provisions and energy payments in lieu of tax provisions which may reduce greenhouse gases. All references to the Code are intended to refer to the Internal Revenue Code of 1986, unless otherwise specified. Summary descriptions of the provisions are provided below and the associated revenue effects are shown in Table 5. A tax expenditure is an exception to baseline provisions of the tax structure that usually results in a reduction in the amount of tax owed. In addition to categories of tax expenditures described in previous Federal climate change expenditures reports, this report contains estimated payments from the Department of the Treasury authorized by Section 1603 of the American Recovery and Reinvestment Act.

All tax expenditure estimates presented here were based upon current tax law enacted as of December 31, 2012. Expired or repealed provisions are not listed if their revenue effects result only from taxpayer activity occurring before fiscal year 2012. Tax expenditure information can also be found in *Analytical Perspectives, Budget of the United States Government*, Fiscal Year 2014, Chapter 16.[5]

Energy production credit – The Code provides a credit for certain electricity produced from wind energy, biomass, geothermal energy, solar energy, small irrigation power, municipal solid waste, qualified hydropower production, or marine and hydrokinetic renewable energy, and sold to an unrelated party.

Energy investment credit — The Code provides credits for investments in solar and geothermal energy property, qualified fuel cell property, qualified microturbine property, geothermal heat pumps, qualified small wind property and combined heat and power property. Owners of renewable power facilities that qualify for the energy production credit may instead elect to take an energy investment credit.

Credit for alternative motor vehicles and refueling property – The Code allows a number of credits for certain types of vehicles and property. These are available for alternative fuel vehicle refueling property, fuel cell vehicles and plug-in electric drive motor vehicles.

Table 5. Energy Tax Provisions That May Reduce Greenhouse Gases (Revenue effect in millions of dollars)

	2012	2013	2014	2015	2016	2017	2018	2014-2018
Energy Production Credit (without coal)[1]	1,452	1,719	1,759	1,719	1,629	1,428	1,088	7,622
Energy Investment Credit[2]	1,040	1,270	1,360	1,670	1,880	1,110	240	6,260
Tax credit for alternative motor vehicles and refueling property[3]	100	180	260	400	610	670	500	2,440
Exclusion of utility conservation subsidies	270	250	250	250	250	250	240	1,240
Credit for holding clean renewable energy bonds	70	70	70	70	70	70	70	350
Allowance of deduction for certain energy efficient commercial building property	70	70	40	20	0	0	-20	40
Credit for construction of new energy efficient homes	70	40	20	0	0	0	0	20
Credit for energy efficiency improvements to existing homes	780	0	0	0	0	0	0	---
Credit for energy efficient appliances	210	300	130	120	100	0	0	350
Credit for residential energy efficient properties[4]	910	1,010	1,140	1,270	1,420	600	0	4,430
Qualified energy conservation bonds	20	30	30	30	30	30	30	150
Industrial CO2 capture and sequestration tax credit	60	60	70	80	110	210	160	630
Tax Provisions Subtotal	*5,052*	*4,999*	*5,129*	*5,629*	*6,099*	*4,368*	*2,308*	*23,532*
Energy Payments in lieu of energy investment credit[2,5]	*5,080*	*8,080*	*4,710*	*2,520*	*1,580*	*330*	*0*	*9,140*

	2012	2013	2014	2015	2016	2017	2018	2014-2018
Tax Provisions plus Energy Payments Total	**10,132**	**13,079**	**9,839**	**8,149**	**7,679**	**4,698**	**2,308**	**32,672**

[1] Estimates of revenue loss from coal provisions have been removed from the tax expenditure estimate in the budget. In previous years, the Expenditures Report cited the New Technology Credit.

[2] In previous years the Energy Investment Credit was contained within the New Technology Credit. The Energy Investment Credit also includes the business installation of fuel cells, which was an independent entry in tables from previous years. These estimates do not exclude microturbine credits which were removed in previous expenditures reports, however the estimates are expected to be too small to affect these figures which are rounded to the nearest $10 million.

[3] In previous reports the tax credit for alternative motor vehicles and refueling property was referred to as the tax credit and deduction for clean-burning vehicles.

[4] In previous years the credit for residential energy efficient property was referred to as the credit for residential purchases/installations of solar and fuel cells.

[5] Firms can take an energy payment in lieu of the energy investment credit for facilities placed in service in 2009, 2010, or 2011 or whose construction commenced in 2009, 2010, or 2011. The payments are considered outlays and are direct substitutes for the energy tax provisions.

Exclusion of utility conservation subsidies – In certain circumstances, public utilities offer rate subsidies to non-business customers who invest in energy conservation measures.

Credit for holding clean renewable energy bonds – The Code provides for the issuance of Clean Renewable Energy Bonds which entitles the bond holder to a Federal income tax credit in lieu of interest. The limit on the volume issued in 2009-2010 is $2.4 billion.

Allowance of deduction for certain energy efficient commercial building property – The Code allows a deduction, per square foot, for certain energy efficient commercial buildings property installed on or in a commercial building.

Credit for construction of new energy efficient homes – The Code allows contractors a tax credit of $2,000 for the construction of a qualified new energy-efficient home with an annual level of heating and cooling energy consumption at least 50 percent below a reference energy standard. The Code also allows a tax credit of $1,000 for the construction of a qualified new energy-efficient manufactured home with an annual level of heating and cooling energy consumption at least 30 percent below a reference energy standard.

Credit for energy efficiency improvements to existing homes – The Code provides an investment tax credit for expenditures made on insulation, exterior windows (including skylights), exterior doors, and metal or asphalt roofs with appropriate pigmented coatings or cooling granules that improve the energy efficiency of a home and meet certain standards. The Code also provides a credit for purchases of advanced main air circulating fans, natural gas, propane, or oil furnaces or hot water boilers, and other qualified energy efficient property.

Credit for residential energy efficient property – The Code provides an investment tax credit for expenditures made on solar electric property, solar hot water heaters, fuel cells, small wind turbines, and geothermal heat pumps for use in a residence.

Credit for energy efficient appliances – The Code provides tax credits for the manufacture of energy efficient dishwashers, clothes washers, and refrigerators. The amount of the tax credit depends on the energy efficiency of the appliance.

Advanced energy property credit – The Code provides a 30 percent investment credit for property used in a qualified advanced energy manufacturing project. The Treasury Department may award up to $2.3 billion in tax credits for qualified investments.

Credit for qualified energy conservation bonds — The Code provides for the issuance of energy conservation bonds which entitle the bond holder to a Federal income tax credit in lieu of interest. The limit on the volume issued in 2009 is $3.2 billion.

Industrial CO2 capture and sequestration tax credit — The Code allows a credit of $20 per metric ton for qualified carbon dioxide captured at a qualified facility and disposed of in secure geological sequestration. The Code also allows a credit of $10 per metric ton of qualified carbon dioxide that is captured at a qualified facility and used as a tertiary injectant in a qualified enhanced oil or natural gas recovery project.

Energy payments in lieu of energy investment credit — Section 1603 of the American Recovery and Reinvestment Tax Act of 2009 (Section 1603) authorizes the Treasury Department to make payments to persons who place in service specified energy property in 2009, 2010, or 2011 or whose construction commenced in 2009, 2010, or 2011. Firms can take an energy payment in lieu of the energy production credit or the energy investment credit.

6. Climate Adaptation, Preparedness, and Resilience

Climate change is a complex, interdisciplinary issue with the potential to affect nearly every sector and level of governmental operations. Across the United States and the world, climate change is already affecting communities, livelihoods, and the environment. To address these challenges and ensure the nation is prepared and resilient to the impacts of climate change, in 2009, the Administration convened the Interagency Climate Change Adaptation Task Force, co-chaired by the Council on Environmental Quality (CEQ), the Office of Science and Technology Policy (OSTP), and the National Oceanic and Atmospheric Administration (NOAA), and including representatives from more than 20 Federal agencies. In addition, on October 5, 2009, President Obama signed an Executive Order directing the Task Force to develop recommendations for how the Federal Government can strengthen policies and programs to better prepare the nation to adapt to the impacts of climate change. In its 2010 Progress Report, the Task Force called on Federal agencies to demonstrate leadership on climate change adaptation. Rising sea levels, drought, extreme weather events, loss of land and sea ice, and other climate-related impacts threaten communities, ecosystems, and Federal services and assets. The 2010 Task Force Report determined that the Federal Government has a responsibility to safeguard Federal services and resources and to help states, tribes, and communities manage climate- related risks by improving access to climate information, enhancing coordination and capacity, and leading and supporting actions that reduce vulnerability and increase resilience. In response, Federal agencies are taking steps to prepare the nation for the impacts of climate change and are making significant progress. These actions are outlined in agencies' first ever Climate Change Adaptation Plans, which were released in February 2013 as part of the annual Strategic

Sustainability Planning Process. These plans outline initiatives to reduce the vulnerability of Federal programs, assets, and investments to the impacts of climate change, such as sea level rise or more frequent or severe extreme weather. Agency adaptation plans highlight actions to plan for and address these impacts in their programs and operations, and protect taxpayer investments.

Table 6. Natural Resources Adaptation (Budget authority in millions of dollars)[1]

Natural Resources Adaptation[1]	FY 2012 Enacted Budget Authority	FY 2013 Enacted Budget Authority	FY 2013 Current Budget Authority[3]	FY 2014 Proposed Budget Authority	Proposed Change in Budget Authority 2013-2014
Department of Interior					
National Park Service Operation of the National Park Service	3	3	3	9	+6
Fish and Wildlife Service—Resource Management	60	67	64	65	-2
Bureau of Land Management Management of Lands and Resources	18	18	16	21	+3
Bureau of Indian Affairs Operation of Indian Programs[2]	0	1	1	10	+9
Bureau of Reclamation Cooperative Landscape Conservation	7	6	6	5	-1
Total – Natural Resources Adaptation	**88**	**95**	**90**	**110**	**+15**

[1] All data supersede numbers released with the 2014 President's Budget. Budget Authority provided in millions of dollars. Discrepancies resulted from rounding and improved estimates. Funding in the table does not include USGS climate change adaptation research, which is captured within USGCRP totals.

[2] BIA activities include: assisting Tribes and Alaska Natives with land and resource management, and adaptive management strategies to deal with the effects of climate change they are experiencing or expect to experience; providing climate change funds to Tribes for mitigation and adaptation projects that are deemed high priority; and provide climate change funds for Tribes to actively engage and

participate in the Climate Science Centers, LCCs, and the many other climate change implementation projects that require tribal input.

[3] Current Budget Authority for FY 2013 throughout this document reflects the amount the program has available for the year calculated as the appropriated amount (as reported in the FY 2013 Enacted column) minus the reductions pursuant to the Budget Control Act of 2011 (P.L. 112-25) sequestration order issued on March 1, 2013, and accounting for any known and applicable reprogrammings, transfers, or other related adjustments. Estimates are current as of June 21, 2013 and are subject to change.

Agencies are also developing collaborative approaches within the government to build coordinated and comprehensive responses to the impacts of climate change in all sectors. The first of these efforts have focused on building the climate preparedness and resilience of natural resources, including oceans and coasts, wildlife, and water resources. Federal agencies worked with stakeholders to develop a National Action Plan for managing freshwater resources in a changing climate to assure adequate water supplies and protect soil and water quality, human health, property, and aquatic ecosystems. Federal agencies also worked with state, tribal, and local representatives to develop a National Fish, Wildlife and Plants Climate Adaptation Strategy, for safeguarding our nation's species and natural resources (http://www.wildlifeadaptationstrategy.gov). The final strategy was released in March 2013.

There are numerous efforts across the Federal Government for preparing and building resilience to the impacts of climate change on various critical sectors, institutions, and agency mission responsibilities. The President's Climate Action Plan highlights many key efforts to advance climate adaptation, preparedness and resilience. Successful efforts to build resiliency and adaptation often involve integrating climate change considerations into existing agency programs, projects, and activities rather than establishing separate and distinct programs. This creates a challenge when attempting to fully account for all adaptation resources. While the Administration continues to develop methodologies to account for a broader suite of adaptation programs across all critical sectors, this report used the following summary of Department of the Interior activities designed to promote adaptation as an example of one agency's efforts in this area. These Department of the Interior activities also reflect a variety of interagency efforts to address key adaptation challenges that cut across the jurisdictions and missions of individual Federal agencies, and affect fresh water, oceans and coasts, and fish, wildlife and plants.

An early example of agency efforts to promote climate preparedness and resilience is the work of the Department of the Interior to gain effective and broad collaboration to determine the causes and implement changes to reduce climate impacts to lands, waters, natural and cultural resources.

A key component of this initiative is the development of a network of Landscape Conservation Cooperatives (LCCs), which are applied conservation science partnerships that provide scientific and technical support for spatially-explicit conservation goals and for integrated, adaptive management actions at landscape scales. LCCs are composed of and depend on other Federal agencies, tribal, local and State partners, and the public in crafting practical, landscape-level strategies for managing climate change impacts in coordination with the Department's Climate Science Centers (CSCs). The focus of the CSCs includes impacts of climate change on fish, wildlife, and habitats, including wildlife migration patterns, wildfire risk, drought, or invasive species that typically extend beyond the borders of any particular Federal or Tribal land holding.

With resident staff and through connections with partners, LCCs develop, test, implement, and monitor conservation strategies that respond to the dynamic landscape changes resulting from climate change. The LCCs facilitate broad availability of data, modeling, and tools to land managers that allow them to analyze and model trends in species and habitat changes. LCCs also support improved management of water resources, historical and cultural resources, and resources that are needed by Indian Tribes and Alaska Natives.

The FY 2014 President's Budget continues support for cooperative landscape conservation in the face of climate change and other environmental stressors. The National Park Service supports managers with the tools to inventory changes and adapt management practices with a $6 million increase. The Bureau of Indian Affairs provides a $9 million increase to better integrate climate adaptation work on trust land and to support Tribal participation with the CSC and the Landscape Conservation Cooperatives, including the use of traditional ecological knowledge in adaptation management. These funding levels provide the critical science to support Interior's $110 million Climate Change Wildlife Adaptation initiative. The Bureau of Reclamation continues to support cooperative landscape conservation, focusing on climate change implications for water resources management and addressing the Department's Priority Goal for Climate Change through vulnerability assessments, adaptation actions, and development of improved assessment tools through collaboration with CSCs and other climate science entities.

APPENDIX. ACCOUNTING OF FEDERAL CLIMATE CHANGE EXPENDITURES BY AGENCY

The following is a listing of Federal climate change expenditures by agency and by line item in the President's 2014 Budget Appendix. Budget Appendix line items show account level data and may not reflect sub-account level climate change information. The data in this table may be subsets of an account.

Table 7. Climate Change Expenditures by Agency Details by Agency/Account (Budget authority in millions of dollars)[1]

Climate Change Expenditures by Agency	FY 2012 Enacted Budget Authority	FY 2013 Enacted Budget Authority	FY 2013 Current Budget Authority[2]	FY 2014 Proposed Budget Authority	Change inBudget Authority 2013-2014
Department of Agriculture					
Global Change Research Program					
Agricultural Research Service	36	36	38	52	+16
National Institute of Food and Agriculture	50	40	40	43	+3
Economic Research Service	2	2	2	2	---
Forest Service –Forest and Rangeland Research	26	25	25	28	+3
National Agricultural Statistics Service	1	1	1	1	---
Natural Resources Conservation Services	1	1	1	1	---
USDA-GCRP Subtotal	**116**	**104**	**106**	**126**	**22**
Clean Energy Technology					
Natural Resources Conservation Service – Conservation Operations	6	6	0	4	-2
Agricultural Research Service –Salaries and Expenses	33	32	32	39	+7
National Institute of Food and Agriculture –Research and Education Activities	31	57	56	51	-5
Forest Service – Commercialization/Renewable Energy	26	23	23	28	+5

Table 7. (Continued)

Climate Change Expenditures by Agency	FY 2012 Enacted Budget Authority	FY 2013 Enacted Budget Authority	FY 2013 Current Budget Authority[2]	FY 2014 Proposed Budget Authority	Change inBudget Authority 2013-2014
Rural Business Service –Value Added Producer Grants (Cooperative Development Grants)	1	2	1	1	-1
Rural Business Service –Rural Energy for America Program	3	3	3	20	+16
Rural Business Cooperative Service –Guaranteed Business and Industry Loans	4	6	5	6	---
Rural Business Cooperative Service –Rural Economic Development Loans	0	0	0	0	---
Economic Research Service	2	2	2	2	---
Office of the Chief Economist –Salaries and Expenses	4	3	3	4	+1
Rural Utilities Service –High Cost Energy Grants	4	4	4	0	-4
2008 Farm Bill, Mandatory Funding					
Rural Business Service –Rural Energy for America	22	0	0	70	+70
National Institute of Food and Agriculture –Biomass Research and Development	40	0	0	26	+26
Farm Service Agency – Biomass Crop Assistance Program	17	0	0	0	---
Farm Service Agency – Commodity Credit Corporation	0	170	161	0	-170
Natural Resources Conservation Service –Farm Security and Rural Investment Programs	16	14	14	14	---
Rural Business Service – Bioenergy Program for Advanced Biofuels	65	0	0	0	---
Subtotal -mandatory funding	*160*	*184*	*175*	*110*	*-74*
Subtotal -discretionary funding	*116*	*138*	*130*	*155*	*+18*
USDA- Clean Energy Subtotal	**275**	**322**	**305**	**265**	**-57**

Climate Change Expenditures by Agency	FY 2012 Enacted Budget Authority	FY 2013 Enacted Budget Authority	FY 2013 Current Budget Authority[2]	FY 2014 Proposed Budget Authority	Change inBudget Authority 2013-2014
International Assistance					
Forest Service-Forest and Rangeland Research	**3**	**3**	**3**	**1**	**-2**
Total-USDA	**394**	**429**	**414**	**392**	**-37**
Department of Commerce					
Global Change Research Program					
National Oceanic and Atmospheric Administration – Operations, Research, and Facilities	245	247	233	307	+60
National Oceanic and Atmospheric Administration – Procurement, Acquisition, and Construction	69	64	64	59	-5
National Institute of Standards and Technology (NIST)	5	5	5	5	---
DOC-GCRP Subtotal	**319**	**316**	**302**	**371**	**+55**
Clean Energy Technology					
National Institute of Standards and Technology (NIST) – Scientific and Technological Research and Services	40	40	40	40	---
National Oceanic and Atmospheric Administration Operations, Research and Facilities	0	0	0	3	+3
DOC-Clean Energy Subtotal	**40**	**40**	**40**	**43**	+3
Total-Department of Commerce	**359**	**356**	**342**	**414**	**+58**
Department of Defense					
Clean Energy Technology					
Research, Development, Test and Evaluation, Army	32	29	29	32	+2
Research, Development, Test and Evaluation, Navy	231	186	176	226	+40
Research, Development, Test and Evaluation, Air Force	118	203	190	153	-50
Research, Development, Test and Evaluation, Defense Wide[5]	101	46	42	46	---
Total-Department of Defense	**481**	**465**	**437**	**457**	**-8**

Table 7. (Continued)

Climate Change Expenditures by Agency	FY 2012 Enacted Budget Authority	FY 2013 Enacted Budget Authority	FY 2013 Current Budget Authority[2]	FY 2014 Proposed Budget Authority	Change inBudget Authority 2013-2014
Department of Energy					
Global Change Research Program					
Science –Biological & Environmental Research	**211**	**213**	**209**	**220**	**+7**
Clean Energy Technology					
Energy Efficiency and Renewable Energy	1,819	1,810	1,719	2,788	+978
Electricity Delivery and Energy Reliability	133	133	126	153	+20
Nuclear Energy	772	765	723	733	-32
Fossil Energy R&D –Carbon Capture and Storage (CCS) and Power Systems	472	446	425	375	-71
Science –Fusion, Sequestration, and Hydrogen	902	924	883	1,067	+143
Energy Transformation Acceleration Fund –Advance Research Projects Agency-Energy (ARPA-E)	275	264	251	379	+114
Bonneville Power Administration Fund	15	17	17	17	---
Race to the Top for Energy Efficiency and Grid Modernization	0	0	0	200	+200
HomeStar	0	0	0	300	+300
Energy Security Trust	0	0	0	200	+200
DOE-Clean Energy Subtotal	**4,388**	**4,359**	**4,144**	**6,212**	**+1,853**
International Assistance					
Energy Efficiency and Renewable Energy	9	9	9	9	---
Fossil Energy R&D –Carbon Capture and Storage (CCS) and Power Systems	3	3	3	3	---
Science	1	1	1	1	---
DOE-International Assistance Subtotal	**13**	**13**	**13**	**13**	**---**
Adjustments for programs included in multiple categories --DOE	*-13*	*-13*	*-13*	*-13*	---
Total-DOE	**4,599**	**4,572**	**4,353**	**6,432**	**+1,860**

Climate Change Expenditures by Agency	FY 2012 Enacted Budget Authority	FY 2013 Enacted Budget Authority	FY 2013 Current Budget Authority[2]	FY 2014 Proposed Budget Authority	Change inBudget Authority 2013-2014
Department of Health and Human Services					
Global Change Research Program					
Centers for Disease Control and Prevention	6	7	7	7	---
National Institutes of Health	8	8	8	8	---
HHS- GCRP Subtotal	14	15	14	15	---
Total-HHS	**14**	**15**	**14**	**15**	**---**
Department of the Interior					
Global Change Research Program					
U.S. Geological Survey – Surveys, Investigations, and Research	**59**	**58**	**55**	**72**	**+14**
Natural Resources Adaptation					
National Park Service – Operation of the National Park Service	3	3	3	9	+6
Fish and Wildlife Service – Resource Management	60	67	64	65	-2
Bureau of Land Management –Management of Lands and Resources	18	18	16	21	+3
Bureau of Indian Affairs – Operation of Indian Programs[2]	0	1	1	10	+9
Bureau of Reclamation – Cooperative Landscape Conservation	7	6	6	5	-1
DOI-Natural Resources Adaptation Subtotal	88	95	90	110	+15
Total-DOI	**147**	**153**	**145**	**182**	**+29**
Department of State					
Global Change Research Program					
Other-non-add	*3*	*3*	*3*	*3*	---
International Assistance					
Diplomatic and Consular Affairs	0	0	0	0	---
Economic Support Fund	96	96	91	94	-3
International Organizations and Programs	37	37	35	39	+2
State-International Assistance Subtotal	**133**	**132**	**126**	**133**	**+1**

Table 7. (Continued)

Climate Change Expenditures by Agency	FY 2012 Enacted Budget Authority	FY 2013 Enacted Budget Authority	FY 2013 Current Budget Authority[2]	FY 2014 Proposed Budget Authority	Change inBudget Authority 2013-2014
Total-State	**133**	**132**	**126**	**133**	**+1**
Department of Transportation					
Global Change Research Program					
Federal Highway Administration –Federal-Aid Highways	0	0	0	0	---
Federal Aviation Administration – Research, Engineering, and Development	1	1	1	1	---
Federal Transit Administration -Research and University Research Centers	0	0	0	0	---
DOT-GCRP Subtotal	**1**	**1**	**1**	**1**	**---**
Clean Energy Technology					
National Highway Traffic Safety Administration	10	10	8	11	+1
Research and Innovative Technology Administration – Research and Development	1	1	1	1	---
Federal Aviation Administration -Research, Engineering, and Development	21	17	20	18	+1
Federal Aviation Administration -Facilities and Equipment	7	5	4	5	+1
Federal Transit Administration - Research and University Research Centers and Formula and Bus Grants	52	23	22	15	-8
Federal Railroad Association - Railroad Research and Development	1	2	1	3	+1
DOT-Clean Energy Subtotal	**91**	**57**	**56**	**52**	**-5**
Total-DOT	**92**	**58**	**57**	**53**	**-5**
Department of the Treasury					
International Assistance					
Debt Restructuring –Tropical Forestry Conservation	12	12	11	0	-12

Climate Change Expenditures by Agency	FY 2012 Enacted Budget Authority	FY 2013 Enacted Budget Authority	FY 2013 Current Budget Authority[2]	FY 2014 Proposed Budget Authority	Change inBudget Authority 2013-2014
Global Environment Facility	60	65	62	72	+7
Clean Technology Fund	230	185	175	216	+31
Strategic Climate Fund	75	50	47	68	+18
Total-Treasury	**377**	**311**	**296**	**356**	**+44**
Environmental Protection Agency					
Global Change Research Program					
Science and Technology	**18**	**19**	**17**	**20**	**+1**
Clean Energy Technology					
Environmental Programs and Management	99	99	95	106	+7
Science and Technology	18	17	16	10	-7
EPA-Clean Energy Subtotal	**117**	**116**	**111**	**115**	---
International Assistance					
Environmental Programs and Management	**18**	**18**	**16**	**19**	**---**
Adjustments for programs included in multiple categories--EPA	*-9*	*-9*	*-7*	*-9*	---
Total-EPA	**144**	**144**	**137**	**145**	**+1**
Millennium Challenge Corporation					
International Assistance					
Millennium Challenge Corporation	**41**	**0**	**0**	**0**	**---**
Total-MCC	**41**	**0**	**0**	**0**	**---**
National Aeronautics and Space Administration					
Global Change Research Program					
Science	**1,390**	**1,444**	**1,428**	**1,493**	**+49**
Clean Energy Technology					
Aeronautics	259	262	255	284	+22
Exploration	9	7	6	9	+1
Space Technology	28	15	15	28	+14
NASA-Clean Energy Subtotal	**296**	**284**	**276**	**321**	**+37**
International Assistance					
Science	**3**	**3**	**3**	**3**	**---**
Adjustments for programs included in multiple categories--NASA	*-3*	*-3*	*-3*	*-3*	---
Total-NASA	**1,686**	**1,728**	**1,704**	**1,814**	**+86**

Table 7. (Continued)

Climate Change Expenditures by Agency	FY 2012 Enacted Budget Authority	FY 2013 Enacted Budget Authority	FY 2013 Current Budget Authority[2]	FY 2014 Proposed Budget Authority	Change inBudget Authority 2013-2014
National Science Foundation					
Global Change Research Program					
Research and Related Activities	**333**	**328**	**316**	**326**	**-2**
Clean Energy Technology					
Research and Related Activities	**341**	**352**	**346**	**372**	**+20**
International Assistance					
Research and Related Activities	**6**	**6**	**6**	**3**	**-3**
Total-NSF	**680**	**686**	**668**	**701**	**+15**
Nuclear Regulatory Commission					
Clean Energy Technology					
Salaries and Expenses	**83**	**82**	**57**	**86**	+4
Total-NRC	**83**	**82**	**57**	**86**	+4
Smithsonian Institution					
Global Change Research Program					
Salaries and Expenses	8	8	8	8	---
Total-Smithsonian	**8**	**8**	**8**	**8**	---
Tennessee Valley Authority					
Clean Energy Technology					
Tennessee Valley Authority Fund	**9**	**11**	**11**	**10**	-1
Total-TVA	**9**	**11**	**11**	**10**	-1
US Trade and Development Agency					
International Assistance					
Trade and Development Agency	**16**	**16**	**0**	**18**	**+1**
Total-TDA	**16**	**16**	**0**	**18**	**+1**
U.S. Agency for International Development					
Global Change Research Program					
Development Assistance-non-add	11	11	11	14	+3
International Assistance					
Assistance for Europe, Eurasia, and Central Asia	15	0	0	0	---
Development Assistance	322	322	308	317	-5
Economic Support Fund	12	28	27	32	+4

Climate Change Expenditures by Agency	FY 2012 Enacted Budget Authority	FY 2013 Enacted Budget Authority	FY 2013 Current Budget Authority[2]	FY 2014 Proposed Budget Authority	Change inBudget Authority 2013-2014
International Disaster Assistance	0	0	0	0	---
USAID-International Assistance Subtotal	**348**	**350**	**334**	**349**	**-1**
Total-USAID	**348**	**350**	**334**	**349**	**-1**
Total All Agencies[1]	**9,649**	**9,519**	**9,116**	**11,569**	**+2,051**
Energy Tax Provisions That May Reduce Greenhouse Gases	5,052	4,999	4,999	5,129	+130
Energy Payments in lieu of energy investment credit	5,080	8,080	8,080	4,710	-3,370
Total All Agencies + Tax Provisions	**19,781**	**22,598**	**22,195**	**21,408**	**-1,189**

[1] Totals may not sum due to rounding.

[2] Current Budget Authority for FY 2013 throughout this document reflects the amount the program has available for the year calculated as the appropriated amount (as reported in the FY 2013 Enacted column) minus the reductions pursuant to the Budget Control Act of 2011 (P.L. 112-25) sequestration order issued on March 1, 2013, and accounting for any known and applicable reprogrammings, transfers, or other related adjustments. Estimates are current as of June 21, 2013 and are subject to change.

End Notes

[1] http://www.globalchange.gov/about

[2] www.wildlifeadaptationstrategy.gov

[3] Summary of the "whole of government" U.S. International Climate Change Financing is available at http://www.state.gov/documents/organization/201130.pdf

[4] Estimates of development finance and export credit agencies' international climate investments are based on an initial review of planned projects, and in some cases the final review of activities, after their implementation, may change the accounting of timing and scale of financing. http://www.state.gov/documents/organization/201130.pdf

[5] Several temporary provisions, including the energy production tax credit, the energy investment credit, the credit for refueling property, the credit for energy efficient improvements to existing homes, the credit for construction of new energy efficient homes, and the credit for energy efficient appliances, were extended or modified under the American Taxpayer Relief Act of 2012. The tax expenditure estimates in this report do not reflect the extension of these tax incentives.

In: Federal Climate Change Funding
Editor: Sandrin Toullart

ISBN: 978-1-62948-554-6

Chapter 3

THE GLOBAL CLIMATE CHANGE INITIATIVE (GCCI): BUDGET AUTHORITY AND REQUEST, FY2010-FY2014*

Richard K. Lattanzio

SUMMARY

The United States supports international financial assistance for global climate change initiatives in developing countries. Under the Obama Administration, this assistance has been articulated primarily as the Global Climate Change Initiative (GCCI), a platform within the President's 2010 Policy Directive on Global Development. The GCCI aims to integrate climate change considerations into U.S. foreign assistance through a range of bilateral, multilateral, and private sector mechanisms to promote sustainable and climate-resilient societies, foster low-carbon growth, and reduce emissions from deforestation and land degradation. The GCCI is implemented through programs at three "core" agencies: the Department of State, the Department of the Treasury, and the U.S. Agency for International Development (USAID). Most GCCI activities at USAID are implemented through the agency's bilateral development assistance programs. Many of the GCCI activities at the Department of State and the Department of the Treasury are implemented

* This is an edited, reformatted and augmented version of a Congressional Research Service publication, CRS Report for Congress R41845, prepared for Members and Committees of Congress, from www.crs.gov, dated May 28, 2013.

through international organizations, including the United Nations Framework Convention on Climate Change's Least Developed Country Fund and Special Climate Change Fund, as well as multilateral financial institutions such as the Global Environment Facility, the Clean Technology Fund, and the Strategic Climate Fund. The GCCI is funded through the Administration's Executive Budget, Function 150 account, for State, Foreign Operations, and Related Programs.

Congress is responsible for several activities in regard to the GCCI, including (1) authorizing periodic appropriations for federal agency programs and multilateral fund contributions, (2) enacting those appropriations, (3) providing guidance to the agencies, and (4) overseeing U.S. interests in the programs and the multilateral funds. Recent budget authority for the GCCI was $323 million in FY2009, $945 million in FY2010, $819 million in FY2011, and $858 million in FY2012, and has been enacted through legislation including the Omnibus Appropriations Act, 2009 (H.R. 1105; P.L. 111-8); the Consolidated Appropriations Act, 2010 (H.R. 3288; P.L. 111- 117); the Supplemental Appropriations Act, 2010 (H.R. 4899; P.L. 111-212); the Department of Defense and Full-Year Continuing Appropriations Act, 2011 (H.R. 1473; P.L. 112-10); and the Consolidated Appropriations Act, 2012 (H.R. 2055; P.L. 112-74). FY2013 contributions to GCCI programming as provided for in the Consolidated and Further Continuing Appropriations Act, 2013 (H.R. 933; P.L. 113-6), have yet to be fully reported by the agencies. The Administration's FY2014 GCCI budget request is $837 million. Congressional committees of jurisdiction for the GCCI include the U.S. House of Representatives Committees on Foreign Affairs (various subcommittees); Financial Services, Subcommittee on Monetary Policy and Trade; and Appropriations, Subcommittee on State, Foreign Operations, and Related Programs; and the U.S. Senate Committees on Foreign Relations, Subcommittee on International Development and Foreign Assistance, Economic Affairs, and International Environmental Protection; and Appropriations, Subcommittee on State, Foreign Operations, and Related Programs.

As Congress considers potential authorizations and/or appropriations for activities administered through the GCCI, it may have questions concerning U.S. agency initiatives and current bilateral and multilateral programs that address global climate change. Some potential concerns may include cost, purpose, direction, efficiency, and effectiveness, as well as the GCCI's relationship to industry, investment, humanitarian efforts, national security, and international leadership. This report serves as a brief overview of the GCCI and its structure, intents, and funding history.

THE GLOBAL CLIMATE CHANGE INITIATIVE

On September 22, 2010, President Obama signed the Presidential Policy Directive on Global Development.[1] The directive called for the elevation of foreign development assistance as a national priority and outlined an integrated approach to development, diplomacy, and national security. The Global Climate Change Initiative (GCCI)—one of the three main pillars to the 2010 directive[2]—aims to integrate climate change considerations into relevant foreign assistance through a range of bilateral, multilateral, and private mechanisms to promote sustainable and resilient societies, foster low-carbon growth, and reduce emissions from deforestation and land degradation. The GCCI is divided into three main programmatic initiatives, or categories: (1) adaptation assistance, (2) clean energy assistance, and (3) sustainable landscapes assistance.

Adaptation

Adaptation programs aim to assist low-income countries with reducing their vulnerability to climate change impacts and building climate resilience. Bilateral and regional programs at the Department of State and USAID target the more vulnerable countries in Africa, Asia, and Latin America and strive to address climate risks in areas including infrastructure, agriculture, health, and water services; to develop capacity for countries to use the best science and analysis for decision making; and to promote sound governance to carry out these decisions. Multilateral initiatives supported by the United States include the Least Developed Country Fund[3] and the Special Climate Change Fund,[4] which focus on climate resilience and food security provisions in countries with the greatest needs; and the Pilot Program for Climate Resilience,[5] which is tasked with coordinating comprehensive strategies in several of the most vulnerable countries to support actions that respond to the potential risks of a changing climate.

Clean Energy

Clean energy programs aim to reduce greenhouse gas emissions from energy generation and energy use by accelerating the deployment of clean energy technologies, policies, and practices. The United States delivers much of its assistance for clean energy deployment through multilateral trust funds. These funds are primarily housed in international financial institutions (e.g., the World Bank); are currently supported by the financial contributions of donor country governments; and provide financial assistance for projects implemented by a variety of organizations, including U.N. agencies, multilateral development banks, nongovernmental organizations, and national institutions. These funds take advantage of existing large-scale greenhouse gas reduction opportunities and establish investment channels for larger private sector financing.

They include the Clean Technology Fund,[6] which aims to spur large-scale clean energy investments in lower-income countries with rapidly growing emissions; the Global Environment Facility,[7] which provides incremental funding[8] for energy and infrastructure projects that support global environmental benefits; and the Program for Scaling-Up Renewable Energy in Low Income Countries,[9] which endeavors to assist the poorest countries expand energy access and stimulate economic growth through the scaled-up deployment of renewable energy strategies.

Bilateral efforts at the Department of State and U.S. Agency for International Development (USAID) seek to complement the multilateral investments by helping to shape development policy and regulatory environments in the recipient countries.

Sustainable Landscapes

Sustainable landscape programs aim to reduce greenhouse gas emissions from deforestation and forest degradation. Bilateral and regional programs at the Department of State and USAID support country-driven policies for forest governance, forest cover and land use change monitoring systems, law-based resource management and land tenure, and on-the-ground efforts to halt deforestation and foster sustainable forest-based livelihoods. Multilateral initiatives include the Forest Investment Program,[10] which tries to address the circumstances that lead to deforestation and increased greenhouse gas emissions in select lower-income countries by improving regulation and enforcement, mobilizing private financing, and securing the social and economic benefits of sound forest management; and the Global Environment Facility, which provides incremental funding for projects that support global environmental benefits such as biodiversity and sustainable land use.

BUDGET AUTHORITY

The Global Climate Change Initiative is funded through programs at the Department of State, the Department of the Treasury, and USAID (i.e., GCCI "core" agencies). Funds for these programs are appropriated in the Administration's Executive Budget, Function 150 account, for State, Foreign Operations, and Related Programs. Recent trends in the GCCI budget authority show it having remained relatively stable since FY2010,[11] and it currently accounts for approximately 1.4% of total programming in the International Affairs Function 150 account budget request.[12] Recent budget authority for the GCCI was reported as $945 million in FY2010, $819 million in FY2011, and $858 million in FY2012.[13] FY2013 contributions to GCCI programming as provided for in congressional appropriations have yet to be fully reported by the agencies. The Administration's FY2014 GCCI budget request is $837 million. Some additional funds for international climate change financing flow through programs at complementary agencies within the federal government (e.g., the Department of Energy, the Environmental Protection Agency, the Department of Agriculture); however, these allocations are defined outside of the GCCI. Budget authority for GCCI programming in core agencies from FY2010 to FY2014 is represented in *Figure 1*. Funding levels by category and by agency account, in both core and complementary agencies (where available), from FY2010 to FY2014, are presented in *Table A-1* and *Table A-2*, respectively.

Congress is responsible for several activities in regard to the GCCI, including (1) authorizing periodic appropriations for federal agency programs and multilateral fund contributions, (2) enacting those appropriations, (3) providing guidance to the agencies, and (4) overseeing U.S. interests in the programs.

Congressional committees of jurisdiction for international climate change programs at the Department of State, the Department of the Treasury, and USAID include the following:

- the U.S. House of Representatives Committee on Foreign Affairs (various subcommittees);
- the U.S. House of Representatives Committee on Financial Services, Subcommittee on Monetary Policy and Trade;
- the U.S. House of Representatives Committee on Appropriations, Subcommittee on State, Foreign Operations, and Related Programs;

- the U.S. Senate Committee on Foreign Relations, Subcommittee on International Development and Foreign Assistance, Economic Affairs, and International Environmental Protection; and
- the U.S. Senate Committee on Appropriations, Subcommittee on State, Foreign Operations, and Related Programs.

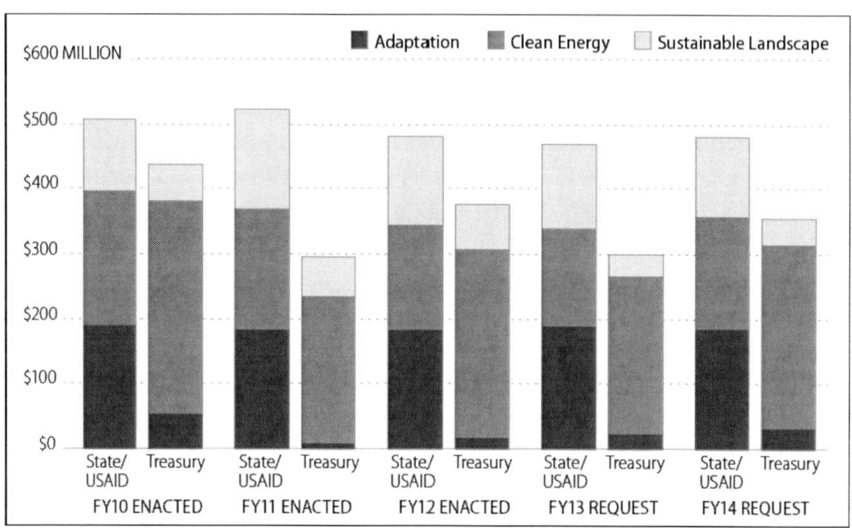

Sources: U.S. Department of State, Congressional Budget Justification, Volume 2, Foreign Operations, Fiscal Years 2010-2014; and the U.S. Administration's foreign assistance website, http://www.foreignassistance.gov/.

Note: See Table A-1 for corresponding figures.

Figure 1. Global Climate Change Initiative: Budget Authority, FY2010-FY2014. (US$ in millions, nominal).

FY2010 Budget Authority

H.R. 3288, the Consolidated Appropriations Act, 2010, was enacted December 16, 2009, as P.L. 111-117. It included appropriations for the Department of State, Foreign Operations, and Related Programs (Division F) that support Global Climate Change Initiative programming. Many GCCI activities are funded by allocations at the sub-account level and were left undefined in P.L. 111- 117. Allocations for FY2010 GCCI sub-account programmatic activities in this report are as reported by agencies on the U.S. Department of State's "Foreign Assistance" website.[14]

Appropriations enacted in P.L. 111-117 related to GCCI activities include the following:

- $86.5 million for the Global Environment Facility (of which the U.S. Department of State estimates that $37 million was allocated toward projects related to global climate change activities, with the remainder allocated to projects related to biodiversity, international waters, ozone protection, organic pollutants, etc.);
- $300 million for the Clean Technology Fund;
- $75 million for the Strategic Climate Fund (including the Pilot Program for Climate Resilience and the Forest Investment Program); and
- larger account level appropriations at the Department of State, USAID, and the Department of the Treasury, including "Development Assistance" at $2,520 million; "Assistance for Europe, Eurasia and Central Asia" at $741.6 million; "Economic Support Fund" at $6,344 million (with another $2,490 million enacted in FY2010 Supplemental Appropriations (H.R. 4899; P.L. 111-212); "Department of the Treasury, Debt Restructuring" at $60 million; and "International Organizations and Programs" at $394 million. From these larger accounts, agencies reported that $508 million in sub-account level bilateral and regional development programming was allocated for GCCI activities.

See the *Appendix* for a breakdown of the FY2010 enacted budget authority.

FY2011 Budget Authority

H.R. 1473, the Department of Defense and Full-Year Continuing Appropriations Act, 2011, was enacted April 15, 2011, as P.L. 112-10. It included appropriations for the Department of State, Foreign Operations, and Related Programs (Title IX) that support Global Climate Change Initiative programming. Many GCCI activities are funded by allocations at the sub-account level and were left undefined in P.L. 112-10. Allocations for FY2011 GCCI sub-account programmatic activities in this report are as reported by agencies on the U.S. Department of State's "Foreign Assistance" website.

Appropriations enacted in P.L. 112-10 related to GCCI activities include the following:

- $89.8 million for the Global Environment Facility (of which the Global Environment Facility reports that approximately 51% was allocated toward projects related to global climate change activities, with the remainder allocated to projects related to biodiversity, international waters, ozone protection, organic pollutants, etc.);
- $184.6 million for the Clean Technology Fund;
- $49.9 million for the Strategic Climate Fund (including the Pilot Program for Climate Resilience, the Forest Investment Program, and the Program for Scaling-Up Renewable Energy in Low Income Countries); and
- larger account level appropriations at the Department of State, USAID, and the Department of the Treasury, including "Development Assistance" at $2,520 million; "Assistance for Europe, Eurasia and Central Asia" at $696 million; "Economic Support Fund" at $5,946 million; "Department of the Treasury, Debt Restructuring" at $50 million; and "International Organizations and Programs" at $354 million.[15] From these larger accounts, agencies reported $523 million in sub-account level bilateral and regional development programming was allocated for GCCI activities.

See the *Appendix* for a breakdown of the FY2011 request and enacted budget authority.

FY2012 Budget Authority

H.R. 2055, the Consolidated Appropriations Act, 2012, was enacted December 23, 2011, as P.L. 112-74. It included appropriations for the Department of State, Foreign Operations, and Related Programs (Division I) that support Global Climate Change Initiative programming. Many GCCI activities are funded by allocations at the sub-account level and were left undefined in P.L. 112- 74. Allocations for FY2012 GCCI sub-account programmatic activities in this report are as reported by agencies on the U.S. Department of State's "Foreign Assistance" website.

Appropriations enacted in P.L. 112-74 related to GCCI activities include the following:

- $89.8 million for the Global Environment Facility (of which the Global Environment Facility reports that approximately 51% was allocated toward projects related to global climate change activities, with the remainder allocated to projects related to biodiversity, international waters, ozone protection, organic pollutants, etc.);
- $184.6 million for the Clean Technology Fund;
- $49.9 million for the Strategic Climate Fund (including the Pilot Program for Climate Resilience, the Forest Investment Program, and the Program for Scaling-Up Renewable Energy in Low Income Countries);
- larger account level appropriations at the Department of State, USAID, and the Department of the Treasury, including "Development Assistance" at $2,520 million; "Assistance for Europe, Eurasia and Central Asia" at $627 million; "Economic Support Fund" at $5,763 million; "Department of the Treasury, Debt Restructuring" at $12 million; and "International Organizations and Programs" at $349 million. From these larger accounts, agencies reported that an estimated $482 million in sub-account level bilateral and regional development programming was allocated for GCCI activities; and
- further, provisions for funding transfers were included, such that in consultation with the Secretary of the Treasury, the Secretary of State may transfer up to $200 million of the funds made available under the Economic Support Fund to funds appropriated under the headings "Multilateral Assistance, Funds Appropriated to the President, International Financial Institutions" for additional payments to such institutions, facilities, and funds. Using these provisions, the Department of State transferred $30 million to the Global Environment Facility (of which $15 million is considered GCCI funding), $45 million to the Clean Technology Fund, and $25 million to the Strategic Climate Fund during FY2012.

See the *Appendix* for a breakdown of the FY2012 request and enacted budget authority.

FY2013 Budget Authority (Pre-sequester)

H.R. 933, the Consolidated and Further Continuing Appropriations Act, 2013, was enacted March 26, 2013, as P.L. 113-6. Under P.L. 113-6, appropriations for the Department of State, Foreign Operations, and Related Programs (Division F, Title VII) that support Global Climate Change Initiative programming were funded through a continuing resolution at the same level as in FY2012, with several changes specified as provisions in the legislation. FY2013-enacted account level estimates are subject to the budget sequestration process as established by the Budget Control Act of 2011 (P.L. 112-25) and the American Taxpayer Relief Act (P.L. 112-240). The total budget impact of sequestration will not be known until the Office of Management and Budget determines the baseline to which it will apply the mandatory reductions. It is expected, however, that the enacted State-Foreign Operations funding for FY2013 will be reduced by approximately 5% through sequestration, and those reductions will be applied at the program, project, and activity level.

As in prior years, many GCCI activities are funded by allocations at the sub-account level and were left undefined in P.L. 113-6. Thus, allocations for FY2013 GCCI sub-account programmatic activities are at the discretion of the agencies, under committee consultation, and have yet to be fully reported by the agencies.

Appropriations enacted in P.L. 113-6 related to GCCI activities (prior to sequestration reductions) include the following:

- $129.4 million for the Global Environment Facility (of which the Global Environment Facility reports that approximately 51% was allocated toward projects related to global climate change activities, with the remainder allocated to projects related to biodiversity, international waters, ozone protection, organic pollutants, etc.);
- $184.6 million for the Clean Technology Fund;
- $49.9 million for the Strategic Climate Fund (including the Pilot Program for Climate Resilience, the Forest Investment Program, and the Program for Scaling-Up Renewable Energy in Low Income Countries);
- larger account level appropriations at the Department of State, USAID, and the Department of the Treasury, including “Development Assistance” at $2,833 million, “Economic Support Fund” at $5,764 million, “Department of the Treasury, Debt Restructuring” at $12 million, and “International Organizations and Programs” at $349 million; and
- further, the provisions for funding transfers, as they existed in P.L. 112-74, were retained in P.L. 113-6.

See the *Appendix* for a breakdown of the FY2013 request and enacted budget authority.

FY2014 Budget Request

The President's FY2013 budget requests for international climate programs at the Department of State, the Department of the Treasury, and USAID would fund near-term climate financing of $836.7 million across three areas:

- adaptation ($220.0 million),
- clean energy ($452.2 million), and
- sustainable landscapes ($164.5 million).

See the *Appendix* for a breakdown of the FY2014 request.

KEY ISSUES FOR CONGRESS

As Congress considers potential authorizations and/or appropriations for initiatives administered through the Department of State, the Department of the Treasury, USAID, and other agencies with international programs, it may have questions concerning the purpose, direction, efficiency, and effectiveness of U.S. agency initiatives and current bilateral and multilateral programs that address global climate change.

Some issues that may raise concerns over providing assistance include the following:

- *Fiscal Constraints.* Budget constraints may lead to questions about sustaining high levels of support for international development assistance in general, and international climate change assistance in particular. The burden is exacerbated during times of economic downturn, when the federal government is hard-pressed to generate fiscal resources to adequately address domestic challenges and maintain basic levels of public services and quality of life. Some have suggested that retaining available funds for immediate domestic priorities, such as fostering renewed economic growth and creating jobs, should take precedence over global concerns for which many Americans feel less urgency and responsibility.

- *Potential for Misuse.* National and international institutions that dispense financial assistance have sometimes been criticized for inefficient and bloated bureaucracies; their lack of transparency about project procurement practices and operating costs; and the proportion of their funds misused or lost through instances of graft, corruption, and other political inefficiencies.[16] Some suggest revisiting operational guidance of these institutions before further appropriations are made.
- *Uncertain Results.* Questions remain regarding the overall effectiveness of international financial assistance in spurring economic development and reform in lower-income countries, and, more specifically, in addressing issues of climate change and the environment. Many studies have examined the effects of international assistance provided to lower-income countries, including both bilateral and multilateral mechanisms, and have returned mixed results, making it difficult to reach firm conclusions that may support continued contributions.
- *Uncertainties in Climate Science.* Prevailing scientific research on the current and future impacts of greenhouse gas emissions on the global climate exhibits varying degrees of analytical uncertainty.
  The lack of definitiveness in some data and in certain model projections has been offered by some as a reason to postpone and/or reconsider international climate change assistance policies and programs.

Some issues that may support providing assistance include the following:

- *Commercial Interests.* Some maintain that international climate change assistance benefits U.S. businesses, as support for low-emission economic growth may increase trade, commerce, and economic activity in the global marketplace for U.S. goods and services. Increased assistance may allow some U.S. industries to make competitive inroads into rapidly expanding markets, improve the advancement and commercialization of U.S. technologies, mobilize greater investment in related domestic sectors, and enhance job creation in the United States.

Decreased funding may cede American influence in global markets to other economic powers still engaged with lower-income countries on environmental and natural resource issues (e.g., the European Union, China).

- *Investment Efficiencies.* Some argue that the costs of responding to tomorrow's climate-related catastrophes, instabilities, conflicts, and technological needs may be significantly higher than the costs of working today to prevent them.
  Some economists note that lower-income countries account for nearly all of the recent growth in global emissions and represent the cheapest near-term opportunity to mitigate GHG pollution as part of a cost-effective solution.
- *Natural Disaster Preparedness.* Some claim that international climate change assistance is a means to support natural disaster preparedness around the globe. Assistance for adaptation activities to help "climate-proof" developing countries' infrastructure and other sectors may help avoid future capital and other losses; minimize the redirection of resources to ad hoc disaster response and urgent humanitarian needs; and avoid chronic humanitarian crises, such as food shortages, particularly for the resource poor in the least developed countries.[17]
- *National Security.* Some defend international climate change assistance as a way to address and mitigate risks to national security. According to a 2008 National Intelligence Assessment, the impacts of global climate change may worsen problems of poverty, social tensions, environmental degradation, and weak political institutions across the developing world.[18]
  Some see international climate change assistance as a means to help make lower-income countries less susceptible to these threats, for the benefit of both the country and the security interests of the United States.
- *International Leadership.* Finally, some see the promotion of international climate change assistance to lower-income countries as a method through which to increase U.S. leadership in global environmental issues.
  Through such leadership, the United States may be able to influence and set important international economic and environmental policies, practices, and standards.[19]

APPENDIX. GCCI BUDGET TABLES

Table A-1. Global Climate Change Initiative, Core Agencies Budget Authority, FY2010-FY2014, by Category (US$ in millions; figures appearing as an asterisk—*—have yet to be reported by the agencies)

Category/ Core Agency[a]	FY2010 Enacted	FY2011 Request	FY2011 Enacted	FY2012 Request	FY2012 Enacted	FY2013 Request	FY2013 Enacted	FY2014 Request
Adaptation[b]	245.8	335.9	194.0	255.0	203.2	215.0	*	220.0
State	68.0	58.5	47.0	48.5	43.0	46.0	*	42.0
USAID	122.8	187.4	137.0	166.5	141.5	144.0	*	144.0
Treasury	55.0	90.0	10.0	40.0	18.7	25.0	25.0	34.0
Clean Energy[c]	531.6	703.5	409.1	662.1	448.3	390.0	*	452.2
State	97.6	66.8	60.5	66.0	68.0	62.5	*	80.0
USAID	108.0	128.7	124.0	129.1	92.0	86.5	*	91.5
Treasury	326.0	508.0	224.6	467.0	288.3	241.0	240.6	280.7
Sustainable	161.9	351.0	215.8	410.5	206.5	164.5	*	164.5
Landscapes[d]								
State	35.9	30.0	17.0	27.5	22.0	12.0	*	10.0
USAID	75.0	175.0	137.4	213.0	115.0	118.5	*	113.5
Treasury	57.0	146.0	61.4	170.0	69.5	34.0	46.0	41.0
Total State	201.5	155.3	124.5	142.0	133.0	120.5	*	132.0
Total USAID	305.7	491.2	398.4	508.6	348.5	349.0	*	349.0
Total	438.0	744.0	296.0	677.0	376.5	300.0	311.6	355.7
Treasury								
Total	945.2	1390.5	818.9	1327.6	858.0	769.5	*	836.7

Sources: U.S. Department of State, Congressional Budget Justification, Volume 2, Foreign Operations, Fiscal Years 2010-2014; and the U.S. Administration's foreign assistance website, http://www.foreignassistance.gov/.

Notes: FY2013 Appropriation totals are prior to sequestration reductions.

[a] GCCI "core agencies" include the Department of State, the Department of the Treasury, and USAID as represented in the Department of State's Executive Budget, Function 150 and Other International Programs (see Table A-2 for a summary of GCCI core and complementary agencies' budget allocations by account).

[b] Adaptation programs aim to assist low-income countries reduce their vulnerability to climate change impacts and build climate resilience.

[c] Clean Energy programs aim to reduce greenhouse gas emissions from energy generation and energy use by accelerating the deployment of clean energy technologies, policies, and practices.

[d] Sustainable Landscape programs aim to reduce greenhouse gas emissions from deforestation and forest degradation.

Table A-2. Global Climate Change Initiative, All Agencies Budget Authority, FY2010-FY2014, by Program (US$ in millions; figures appearing as an asterisk—*—have yet to be reported by the agencies)[a]

Agency/Program/ Account	FY2010 Enacted	FY2011 Enacted	FY2012 Enacted	FY2013 Enacted	FY2014 Request
Department of State[b]	201.5	124.5	133.0	*	120.5
International Organizations—International Conservation Programs (IO&P)[c]	7.5	0.0	0.0	*	0.0
International Organizations—Intergovernmental Panel on Climate Change / UN Framework Convention on Climate Change (IO&P)	13.0	10.0	10.0	*	13.0
International Organizations—Montreal Protocol (IO&P)	23.5	25.5	27.0	*	27.5
Oceans and International Environment and Scientific Affairs (ESF)[d]	145.5	80.0	90.0	*	75.0
Western Hemisphere Regional (ESF)	12.0	9.0	6.0	*	5.0
U.S. Agency for International Development[e]	305.7	398.4	348.5	*	349.0
Bilateral Country (DA,f ESF,[g] AEECA[h])	119.0	212.5	178.0	*	173.1
Africa Regional (DA)	8.5	8.0	8.0	*	5.0
Asia Middle East Regional			2.0		
(DA)	11.5	6.0		*	3.0
Barbados and Eastern			4.5		
Caribbean (DA)	5.0	5.5		*	5.5

Table A-2. (Continued)

Agency/Program/ Account	FY2010 Enacted	FY2011 Enacted	FY2012 Enacted	FY2013 Enacted	FY2014 Request
Bureau for Policy, Planning			1.0		
and Learning (DA)	0.0	0.0		*	1.0
Central Africa Regional (DA)	0.0	5.4	9.4	*	9.4
Central America Regional			6.5		
(DA)	6.0	4.0		*	7.0
Central Asia Regional (ESF)	4.6	1.5	0.0	*	0.0
Democracy, Conflict and Humanitarian Assistance—			0.0		
FEWSNet (DA) Democracy, Conflict and Humanitarian Assistance—	18.0	0.0	8.5	*	0.0
PPM (DA)	2.0	5.0		*	11.0
East Africa Regional (DA)	5.8	8.0	7.0 73.0	*	7.0
Economic Growth, Agriculture and Trade (DA)	76.0	82.0		*	72.0
Eurasia Regional (ESF)	10.0	10.0	3.5	*	3.5
Europe Regional (ESF)	0.0	1.0	1.0	*	1.0
Latin America and Caribbean			5.0		
Regional (DA)	3.5	3.0		*	5.0
Office of Development			0.0		
Partners—Development					
Grants Program (DA)	0.0	0.0		*	0.0
Office of Development			0.0		
Partners—Private Sector					
Alliances (DA)	0.0	0.0		*	0.0
Regional Development			0.0		
Mission for the Pacific (DA)	0.0	0.0		*	0.0
Regional Development			17.0		

Agency/Program/ Account	FY2010 Enacted	FY2011 Enacted	FY2012 Enacted	FY2013 Enacted	FY2014 Request
Mission-Asia (DA)	18.8	23.0		*	20.5
South America Regional (DA)	0.0	4.0	5.6	*	9.0
South Asia Regional (DA)	2.0	2.0	1.5	*	0.0
Southern Africa Regional			7.0		
(DA)	4.2	7.0		*	6.0
USAID Forward: Program			0.0		
Effectiveness Initiatives (DA)	0.0	0.0		*	0.0
West Africa Regional (DA)	10.9	10.5	10.0	*	10.0
Department of the Treasury[i]	432.0	296.0	376.5	311.6	300.0
Tropical Forest Conservation	26.0	16.4	12.0	12.0	0.0
Act (Debt Restructuring)[j]					
Global Environment Facility[k]	37.0	45.0	60.0	65.0	72.0
(IFI)[l]					
Clean Technology Fund[m] (IFI)	300.0	184.6	229.6	184.6	215.7
Strategic Climate Fund: Pilot	55.0	10.0	18.7	25.0	34.0
Program for Climate					
Resilience[n] (IFI)					
Strategic Climate Fund:	20.0	30.0	37.5	12.5	17.0
Forest Investment Program[o]					
(IFI)					
Strategic Climate Fund:	0.0	10.0	18.7	12.5	17.0
Scaling Up Renewable Energy					
Program[p] (IFI)					
Total	945.2	818.9	858.0	*	836.7
(Core Agencies)					
Complementary	360.8	981.1	*	*	*
Agencies[q]					

Table A-2. (Continued)

Agency/Program/ Account	FY2010 Enacted	FY2011 Enacted	FY2012 Enacted	FY2013 Enacted	FY2014 Request
Development Finance and Export Credit Agencies[f]	400.0	1,300.0	*	*	*
Total	1,700.0	3,100.0	*	*	*
(All Agencies)					

Sources: U.S. Department of State, Congressional Budget Justification, Volume 2, Foreign Operations, Fiscal Years 2010-2014; the Department of State's "U.S. Fast Start Climate Finance" website, http://www.state oes/climate/faststart/ index.htm; and the U.S. Administration's foreign assistance website, http://www. foreignassistance.gov/.

Notes: FY2013 Appropriation totals are prior to sequestration reductions.

[a] Figures appearing as an asterisk—*—in this table represent funding estimates that have yet to be allocated and reported fully by the agencies. As of this publication, the Department of State and USAID have yet to report FY2013 estimates, and the complementary agencies and the development finance and export credit agencies have yet to report FY2012 estimates.

[b] U.S. Department of the State GCCI figures not reported as actual are considered estimates and may change as offices update operational budgets and outlays are finalized.

[c] International Organizations and Programs account (IO&P) includes sub-account programmatic activities to support international climate change activities, including contributions to the International Panel on Climate Change, U.N. Framework Convention on Climate Change, and the Montreal Protocol Multilateral Fund.

[d] Economic Support Fund account (ESF) includes sub-account programmatic activities to support international climate change activities, including funding for the Office of Oceans and International Environmental and Scientific Affairs (which contributes funds for adaptation programming at the U.N. Least Developed Countries Fund (LDCF) and the U.N. Special Climate Change Fund (SCCF), clean energy programming at the Major Economies Forum/Clean Energy Ministerial and Climate Renewables and Efficiency Deployment Initiative, and the Global Methane Initiative, and sustainable landscape programming at the Forest Carbon Partnership Facility), and the Office of Western Hemisphere (which contributes to the Energy and Climate Partnership of the Americas).

[e] U.S. Agency for International Development (USAID) GCCI figures not reported as actual are considered estimates and may change as offices update operational budgets and outlays are finalized.

[f] Development Assistance account (DA) includes sub-account programmatic f activities to support USAID bilateral and regional climate change activities.

[g] Economic Support Fund account (ESF) includes sub-account programmatic activities to support USAID/DOS bilateral and regional climate change activities.

[h] Assistance for Europe, Eurasia, and Central Asia account (AEECA) includes sub-account programmatic activities to support USAID bilateral and regional climate change activities. Prior to FY2009, this account was two separate accounts: Assistance for Eastern Europe and the Baltic States and Assistance for the Independent States of the Former Soviet Union.

[i] U.S. Department of the Treasury GCCI figures not reported as actual are considered estimates and may change as offices update operational budgets and outlays are finalized.

[j] Debt Restructuring account (DR) includes sub-account programmatic activities to support international climate change activities in sustainable landscapes programming including funding for the Tropical Forest Conservation Act.

[k] Global Environment Facility (GEF) is a multilateral environmental trust fund that supports climate change activities in clean energy and sustainable landscapes programming. Established in 1991. The United States has contributed funds annually since 1993. Only a portion of GEF funds—as determined through GEF programming decisions—is allocated to climate change activities; the remaining allocation supports other environmental sectors (e.g., biodiversity, oceans, ozone, chemicals). The figures in the above table only present the GEF "climate change" allocation.

[l] International Financial Institutions account (IFI) includes sub-account programmatic activities to support organizations at the World Bank, the regional development banks, and other multilateral institutions.

[m] Clean Technology Fund (CTF) (one of the two World Bank entrusted Climate Investment Funds) is a multilateral environmental trust fund that supports climate change activities in clean energy programming. Established in 2008. The United States has contributed funds annually since 2010.

[n] Strategic Climate Fund (SCF) (one of the two World Bank entrusted Climate Investment Funds): Pilot Program for Climate Resilience (PPCR) is a multilateral environmental trust fund that supports climate change activities in adaptation programming. Established in 2008. The United States has contributed funds annually since 2010.

[o] Strategic Climate Fund (SCF) (one of the two World Bank entrusted Climate Investment Funds): Forest Investment Program (FIP) is a multilateral environmental trust fund that supports climate change activities in sustainable landscapes programming. Established in 2008. The United States has contributed funds annually since 2010.

[p] Strategic Climate Fund (SCF) (one of the two World Bank entrusted Climate Investment Funds): Scaling Up Renewable Energy Program (SREP) is a multilateral environmental trust fund that supports climate change activities in clean energy programming. Established in 2008. The United States did not contribute funds in 2010 because the program was not yet operational.

[q] Complementary agencies (and their programmatic activities) include U.S. Department of Energy (including the China and India Clean Energy Research Centers); U.S. Environmental Protection Agency (including Methane for Markets, International Capacity Building, and contributions to the Multilateral Fund to support the Montreal Protocol on Substances that Deplete the Ozone Layer); U.S. Department of Commerce (including the National Oceanic and Atmospheric Administration's International Research Institute for Climate and Society and the International Trade Administration's activities under the Asia Pacific Partnership); National Science Foundation (including the Regional Institutes for Global Change and the Basic Research to Enable Agricultural Development); U.S. Department of Agriculture (including international programmatic activities in the Forestry Service); National Aeronautics and Space Administration (including the SERVIR initiative); Millennium Challenge Corporation; and U.S. Trade and Development Agency; but do not include Overseas Private Investment Corporation or Export-Import Bank. Figures represented in this table are as reported by the Department of State's "U.S. Fast Start Climate Finance" website, http://www.state.gov/e/ oes/climate/faststart/index.htm. Activities in complementary agencies are not officially defined as GCCI programming.

[r] Development finance and export credit agencies include the Overseas Private Investment Corporation (OPIC) and the Export-Import Bank of the United States (Ex-Im) which use public money to mobilize much larger sums of private investment to be directed at industry sectors (such as climate change mitigation activities) through loans, loan guarantees and insurance for the deployment of clean energy technologies in developing countries. Figures represented in this table are a combination of public and private funding as reported by the Department of State's "U.S. Fast Start Climate Finance" website, http://www.state.gov/e/ oes/climate/faststart/index.htm. Activities in the development finance and export credit agencies are not officially defined as GCCI programming.

End Notes

[1] See White House Fact Sheet: U.S. Global Development Policy at http://www.whitehouse.gov/ the-press-office/2010/ 09/22/fact-sheet-us-global-development-policy.

[2] The GCCI was one of three main initiatives to the 2010 Global Development Policy that also included the Global Health Initiative and Global Food Security.

[3] For more information on this program, see the United Nations Framework Convention on Climate Change website at http://unfccc.int/cooperation_and_support/financial_ mech anism/least_developed_country_fund/items/3660.php.

[4] For more information on this program, see the United Nations Framework Convention on Climate Change website at http://unfccc.int/cooperation_and_support/financial_ mech anism/special_climate_change_fund/items/3657.php.

[5] For more information on this program, see CRS Report R41302, International Climate Change Financing: The Climate Investment Funds (CIF), by Richard K. Lattanzio.

[6] For more information on this program, see CRS Report R41302, International Climate Change Financing: The Climate Investment Funds (CIF), by Richard K. Lattanzio.

[7] For more information on this program, see CRS Report R41165, International Environmental Financing: The Global Environment Facility (GEF), by Richard K. Lattanzio.

[8] "Incremental" funding refers to costs associated with transforming a project with national benefits into one with global environmental benefits. Global Environment Facility (GEF) grants aim to cover the difference or "increment" between a less costly, more polluting option and a costlier, more environmentally sound option. In this way, GEF funding is structured to "supplement" base project funding and provide for the environmental components in national development agendas.

[9] For more information on this program, see CRS Report R41302, International Climate Change Financing: The Climate Investment Funds (CIF), by Richard K. Lattanzio.

[10] For more information on this program, see CRS Report R41302, International Climate Change Financing: The Climate Investment Funds (CIF), by Richard K. Lattanzio.

[11] While the GCCI was instituted as a development initiative during the Obama Administration's first budget submission to Congress for FY2010, the United States has actively contributed congressionally appropriated funds for international climate assistance for many years. However, it should be noted that prior contributions to climate assistance were often accounted for and defined in different ways. Thus, it may be difficult to compare past spending patterns and priorities against those of the current Administration. The Office of Management and Budget (OMB) has reported FY2009 budget authority for activities defined as international climate change assistance for the three core agencies as approximately $323 million. See OMB, Federal Climate Change Expenditure Reports to Congress, June 2010, at http://www.whitehouse.gov/sites/default/files/omb/assets/legislative_reports/FY2011_Climate_Change.pdf; account level budget authority from FY2009 Base Enacted (P.L. 111-8); FY2009 Bridge Enacted (P.L. 110-252); FY2009 Stimulus Enacted (P.L. 111-5); FY2009 Supp. Enacted (P.L. 111-32).

[12] Estimate based on the FY2013 International Affairs, Function 150 account, budget request of $56.2 billion for foreign operations programming at the Departments of State, Treasury, and USAID. Considering that the International Affairs budget generally accounts for approximately 3% of U.S. discretionary spending, the GCCI accounts for less than 0.1% of all U.S. discretionary spending. For more detail on recent Department of State, Foreign Operations, and Related Programs budgets, see CRS Report R43043, The FY2014 State and Foreign Operations Budget Request, by Susan B. Epstein, Marian Leonardo Lawson, and Alex Tiersky.

[13] Beginning with FY2010, the State Department began reporting international climate change assistance on its foreign aid website, ForeignAssistance.gov, at http://www.foreignassistance.gov/InitiativeLanding.aspx. Many GCCI activities are funded at agency sub-account levels, with allocations left to the discretion of the agencies, under congressional consultation. Thus, obligations and/or outlays could change from the enacted budget authority as offices, bureaus, and/or missions update priorities in their operational budgets.

[14] http://www.foreignassistance.gov/.

[15] The larger account figures reflect the 0.2% rescission across all non-defense accounts for FY2011 funds, in accordance with Section 1119(a) of P.L. 112-10.

[16] William Easterly and Tobias Pfutze, "Where Does the Money Go? Best and Worst Practices in Foreign Aid," Journal of Economic Perspectives, vol. 22, no. 2 (Spring 2008). For more on

foreign aid reform, also see CRS Report R40102, Foreign Aid Reform: Studies and Recommendations, by Susan B. Epstein and Matthew C. Weed; and CRS Report R40756, Foreign Aid Reform: Agency Coordination, by Marian Leonardo Lawson and Susan B. Epstein.

[17] Both the World Bank and U.S. Geological Survey estimate that every dollar spent on disaster preparedness saves $7 in disaster response. The World Bank, Natural Disasters: Counting the Cost, March 2, 2004, at http://go.worldbank.org/NQ6J5P2D10.

[18] National Intelligence Council, National Intelligence Assessment on the National Security Implications of Global Climate Change to 2030, Statement for the Record by Dr. Thomas Fingar, Deputy Director of National Intelligence for Analysis, National Intelligence Council, before the U.S. Congress, House Permanent Select Committee on Intelligence & House Select Committee on Energy Independence and Global Warming, June 25, 2008, at http://www.dni.gov/nic/ special_climate2030.html.

[19] For more information on the political issues involved in the international climate change negotiations, see CRS Report R40001, A U.S.-Centric Chronology of the International Climate Change Negotiations, by Jane A. Leggett.

INDEX

#

21st century, 21

A

access, 31, 50, 61, 78
accounting, 6, 15, 18, 22, 27, 36, 43, 54, 63, 73
activity level, 84
adaptation, vii, viii, 2, 3, 14, 15, 18, 19, 23, 24, 25, 29, 30, 31, 44, 48, 49, 50, 51, 56, 61, 62, 63, 64, 77, 85, 87, 92, 93
advancement, 44, 86
aerosols, 31
Africa, 54, 77, 89, 90, 91
agencies, vii, viii, ix, x, 1, 2, 3, 4, 5, 8, 12, 13, 14, 15, 16, 19, 20, 23, 25, 27, 28, 29, 46, 48, 49, 50, 51, 53, 55, 56, 61, 63, 64, 73, 75, 76, 78, 79, 80, 81, 82, 83, 84, 85, 88, 89, 92, 94, 95
agency initiatives, xi, 76, 85
aggregation, 14, 15
Agricultural Research Service, 34, 39, 65
agriculture, 49, 51, 53, 77
Air Force, 40, 67
air quality, 32, 33, 43, 46
Alaska, 62, 64
Alaska Natives, 62, 64
alternative energy, 25
American Recovery and Reinvestment Act, viii, 1, 4, 8, 9, 11, 13, 14, 57
American Recovery and Reinvestment Act of 2009, viii, 1, 4, 8, 9, 11, 13, 14
appropriations, x, 4, 9, 11, 13, 16, 49, 76, 79, 80, 81, 82, 83, 84, 85, 86
Appropriations Act, x, 19, 22, 76, 81, 83
Asia, 23, 54, 77, 89, 91, 94
assessment, 33, 64
assessment tools, 64
assets, 6, 13, 15, 20, 61, 95
atmosphere, 33, 54
authorities, 13
authority, x, 4, 5, 6, 7, 10, 12, 13, 15, 18, 19, 20, 26, 27, 34, 36, 39, 42, 49, 52, 53, 62, 65, 76, 79, 81, 82, 83, 85, 95

B

Barbados, 89
barriers, 45
base, 23, 95
basic research, 28
benefits, 4, 33, 44, 49, 51, 53, 78, 86, 95
BIA, 62
bilateral, 77, 78, 89
biodiversity, 33, 34, 48, 49, 78, 81, 82, 83, 84, 93
biomass, 24, 57
Biomass Crop Assistance Program, 40, 66

blueprint, 38
boilers, 60
bonds, 58, 59, 60
breakdown, 37, 81, 82, 83, 85
budget allocation, 88
budgetary resources, 20
Bureau of Indian Affairs, 62, 64, 69
Bureau of Land Management, 62, 69
businesses, 21, 86
buyers, 51

C

capacity building, 49
carbon, ix, 3, 10, 11, 19, 21, 32, 33, 37, 38, 43, 45, 49, 50, 55, 56, 60, 75, 77
carbon dioxide, 43, 60
carbon emissions, 45
Caribbean, 54, 89, 90
catastrophes, 87
category a, 79
category d, 24
Central Asia, 52, 54, 72, 81, 82, 83, 90, 93
challenges, 19, 25, 32, 33, 44, 46, 61, 63, 85
chemicals, 93
Chile, 51
China, 87, 94
citizens, ix, 22
civil society, 49
clarity, 15, 16
clean air, 32
Clean Air Act, 19
clean energy, 5, 21, 23, 24, 37, 38, 44, 46, 47, 48, 49, 51, 55, 56, 77, 78, 85, 88, 92, 93, 94
Clean Energy Investments, viii, 2, 23
CO2, 58, 60
coal, 47, 54, 58, 59
coatings, 60
collaboration, ix, 2, 19, 64
collaborative approaches, 63
combustion, 43
commerce, 86
commercial, 38, 58, 59
communication, 23, 51
communities, ix, 8, 22, 23, 25, 31, 48, 61
community, ix, 22, 23, 31, 49
comparative advantage, 50
competitiveness, 31
compilation, 19
complement, 50, 55, 78
composition, 33
conflict, 16
congress, 44
Congress, v, vii, viii, ix, x, 1, 2, 4, 5, 6, 7, 8, 9, 11, 12, 16, 18, 19, 20, 21, 22, 25, 29, 54, 75, 76, 79, 85, 95, 96
conservation, 43, 58, 59, 60, 64
conserving, 30, 56
Consolidated Appropriations Act, x, 19, 22, 76, 80, 82
construction, 44, 45, 46, 58, 59, 61, 73
consumers, 38, 44
consumption, 60
cooling, 59, 60
cooperation, 94
coordination, 28, 31, 61, 64
corruption, 86
cost, xi, 45, 47, 76, 87
creativity, 33
CSCs, 29, 64
customers, 59
cycles, 33

D

decision makers, 54
deduction, 58, 59
deforestation, ix, 12, 50, 75, 77, 78, 88
degradation, ix, 12, 34, 75, 77, 78, 88
Department of Agriculture, 9, 10, 12, 29, 34, 39, 43, 53, 65, 79
Department of Commerce, 9, 10, 12, 28, 30, 35, 40, 43, 51, 67
Department of Defense, x, 10, 20, 40, 44, 67, 76, 81
Department of Energy, vii, 1, 4, 9, 10, 11, 12, 28, 30, 35, 37, 41, 44, 52, 55, 68, 79, 94

Department of Health and Human Services, 9, 31, 35, 69
Department of the Interior, 9, 13, 15, 25, 28, 31, 35, 63, 64, 69
Department of the Treasury, x, 9, 12, 50, 52, 53, 57, 70, 75, 79, 81, 82, 83, 84, 85, 88, 91
Department of Transportation, 9, 11, 32, 35, 41, 45, 70
developed countries, 87
developing countries, ix, 11, 34, 48, 49, 50, 54, 55, 75, 87, 94
development assistance, x, 12, 75, 77, 85
development banks, 78, 93
development policy, 78
diplomacy, 77
diplomatic efforts, 50
disaster, 87, 96
distribution, 24
DOC, 30, 44, 67
DOI, 13, 31, 32, 69
donors, 50
DOT, 32, 36, 45, 70
drought, 20, 61, 64

E

early warning, 20
Eastern Europe, 54, 93
ecological restoration, 30
ecological systems, 31
economic activity, 86
economic development, 46, 51, 56, 86
economic downturn, 85
economic growth, 50, 56, 78, 85, 86
economic power, 87
Economic Research Service, 34, 39, 42, 65, 66
economics, 42
ecosystem, 23, 34
education, 23, 33
El Salvador, 54
electricity, 24, 38, 57
emission, ix, 22, 49, 56, 86
energy, vii, viii, ix, 1, 3, 5, 6, 7, 10, 14, 21, 22, 23, 24, 27, 31, 37, 38, 42, 43, 44, 45, 46, 47, 48, 49, 50, 51, 54, 55, 56, 57, 58, 59, 60, 61, 73, 77, 78, 88
energy conservation, 58, 59, 60
energy consumption, 24, 45, 59
energy efficiency, 24, 38, 43, 44, 45, 46, 47, 50, 51, 58, 60
energy security, 37, 44, 54
enforcement, 78
engineering, 28, 30
environment, 8, 25, 37, 47, 61, 86
environmental change, 31, 33
environmental degradation, 87
environmental issues, 44, 87
Environmental Protection Agency, 9, 12, 19, 27, 32, 35, 42, 46, 52, 55, 71, 79, 94
environmental regulations, 55
environmental stress, 34, 64
environmental threats, 31
environments, 49, 78
EPA, 19, 32, 46, 54, 71
equipment, 44
Eurasia, 52, 54, 72, 81, 82, 83, 90, 93
Europe, 52, 54, 72, 81, 82, 83, 90, 93
European Union, 87
evolution, 33
exclusion, 18
Executive Order, 13, 61
exercise, 16
expenditures, viii, 1, 4, 5, 7, 8, 18, 20, 24, 27, 29, 57, 59, 60, 65
expertise, 14, 31, 49, 55
exporters, 51
exports, 51, 55
extreme weather events, 33, 61

F

Farm Bill, 39, 66
farmers, 21, 43
farmland, 43
farms, 30
federal agency, x, 76, 79

federal government, vii, 3, 25, 38, 61, 63, 79, 85
Federal Highway Administration, 35, 36, 70
financial, ix, x, 11, 12, 20, 49, 50, 75, 76, 78, 86, 94
financial resources, 50
fish, 25, 63, 64
Fish and Wildlife Service, 62, 69
flights, 28
food, 48, 77, 87
food security, 48, 77
force, 44
forecasting, 30, 34
foreign aid, viii, 2, 19, 95, 96
foreign assistance, ix, 11, 53, 75, 77, 80, 88, 92
forest management, 30, 78
foundations, 21
freshwater, 63
fuel cell, 45, 57, 59, 60
funding, vii, viii, ix, xi, 1, 2, 3, 4, 5, 6, 8, 9, 10, 11, 13, 14, 15, 16, 19, 20, 22, 25, 28, 29, 36, 37, 38, 40, 42, 43, 48, 54, 55, 64, 66, 76, 78, 83, 84, 87, 92, 93, 94, 95
funds, x, 10, 11, 13, 15, 18, 19, 20, 27, 28, 53, 54, 55, 62, 76, 78, 79, 83, 85, 86, 92, 93, 95

G

GAO, 15, 16, 19, 20
GCCI, v, ix, x, 11, 12, 13, 20, 56, 75, 76, 77, 79, 80, 81, 82, 83, 84, 88, 92, 93, 94, 95
GEF, 53, 93, 95
geothermal heat pump, 57, 60
GHG, vii, ix, 3, 8, 10, 14, 15, 19, 22, 32, 46, 87
GHG emissions, vii, ix, 3, 8, 10, 14, 15, 22, 32, 46
Global Change Research Act, 8, 25
global climate change, vii, ix, xi, 1, 3, 11, 20, 37, 48, 49, 75, 76, 81, 82, 83, 84, 85, 87
Global Climate Change Initiative, v, ix, 5, 11, 12, 13, 20, 56, 75, 77, 79, 80, 81, 82, 84, 88, 89
Global Environment Facility(GEF), x, 12, 50, 52, 54, 71, 76, 78, 81, 82, 83, 84, 91, 93, 95
global markets, 87
global scale, 31, 32
goods and services, 86
governance, 49, 77, 78
governments, 49, 55, 78
grant programs, 24
grants, 15, 43, 95
granules, 60
grasslands, 30
greenhouse, vii, viii, ix, 1, 2, 3, 5, 19, 22, 23, 24, 25, 37, 43, 44, 45, 57, 77, 78, 86, 88
greenhouse gas(GHG), vii, viii, ix, 1, 2, 3, 5, 19, 22, 23, 24, 25, 37, 43, 44, 45, 57, 77, 78, 86, 88
greenhouse gas emissions, 23, 24, 25, 37, 43, 44, 45, 77, 78, 86, 88
greenhouse gases, viii, 2, 19, 57
Gross Domestic Product, 7, 18
growth, ix, 11, 34, 56, 75, 77, 87
guidance, x, 56, 76, 79, 86

H

habitat, 64
habitats, 64
health, 23, 31, 47, 49, 53, 77
health effects, 31
HHS, 69
history, xi, 23, 76
homes, 58, 59, 60, 73
host, 49
House, x, 22, 76, 79, 96
House of Representatives, x, 22, 76, 79
housing, 45
Housing and Urban Development, 45
human, 8, 23, 25, 26, 31, 63
human health, 31, 63
humanitarian crises, 87

hybrid, 45
hydrogen, 45

I

image, 19
impact assessment, 47
improvements, 10, 15, 37, 43, 51, 58, 60, 73
income, viii, 2, 14, 15, 19, 20, 59, 60, 77, 78, 86, 87, 88
income tax, 59, 60
independence, 38, 96
India, 94
industries, 44, 86
industry, xi, 45, 46, 51, 76, 94
inflation, 6, 14
infrastructure, 23, 50, 51, 55, 77, 78, 87
institutions, 25, 50, 63, 78, 83, 86, 87, 93
insulation, 60
international financial institutions, 78, 83, 93
investment(s), xi, 3, 6, 8, 10, 14, 23, 24, 27, 28, 29, 31, 33, 45, 46, 48, 49, 50, 51, 55, 56, 57, 59, 60, 61, 62, 73, 76, 78, 86
investors, 49
irrigation, 51, 57
issues, vii, viii, ix, 2, 3, 15, 19, 47, 85, 86, 87, 96
iteration, 10

J

Japan, 51
job creation, 86
jurisdiction, x, 76, 79

L

land tenure, 78
landscape(s), 30, 48, 54, 56, 64, 77, 78, 85, 92, 93
Latin America, 77, 90
laws, 16, 19
lead, 21, 28, 33, 46, 49, 50, 78, 85
leadership, xi, 23, 30, 37, 45, 48, 49, 51, 61, 76, 87
learning, 56
Least Developed Countries, 92
legislation, x, 4, 16, 19, 76, 84
legislative proposals, ix, 2, 19
Liberia, 54
light, 15
loan guarantees, 43, 49, 94
loans, 43, 49, 94

M

magnitude, 4, 6
majority, vii, 1, 4, 10
management, 31, 43, 49, 54, 62, 64
manufacturing, 38, 43, 44, 60
marketplace, 86
mass, 34
measurement, 43, 44
Mexico, 51
Middle East, 89
migration, 64
military, 44
mission(s), viii, ix, 1, 2, 5, 13, 14, 22, 23, 25, 46, 47, 51, 63, 95
models, 28, 31, 47
Montreal Protocol, 54, 89, 92, 94
Morocco, 54
multilateral, 50, 54, 77, 78, 83, 92, 94
multiplier, 48
municipal solid waste, 57

N

National Aeronautics and Space Administration, 9, 11, 28, 32, 36, 42, 46, 53, 55, 71, 94
National Institutes of Health, 31, 35, 69
National Park Service, 62, 64, 69
national security, xi, 76, 77, 87
natural disaster(s), 28, 87
natural gas, 60
natural hazards, 33
natural resources, 23, 29, 63

Natural Resources Conservation Service, 35, 39, 40, 65, 66
Next Generation Air Transportation System, 46
NextGen, 46
NOAA, 28, 61
NRC, 47, 72
Nuclear Regulatory Commission, 11, 42, 43, 47, 72
nutrient, 43

O

Obama, viii, ix, 2, 3, 11, 19, 21, 22, 38, 75, 95
Obama Administration, viii, ix, 2, 3, 11, 19, 22, 38, 75, 95
oceans, 25, 63, 93
Office of Management and Budget, v, viii, 1, 4, 5, 6, 7, 8, 9, 11, 12, 16, 19, 20, 21, 25, 84, 95
officials, 16
oil, 37, 45, 60
OMB, viii, 1, 4, 5, 7, 8, 10, 11, 13, 15, 16, 18, 19, 20, 25, 95
omission, 15
Omnibus Appropriations Act,, x, 76
open markets, 51
operating costs, 86
operations, ix, 22, 44, 61, 95
opportunities, 45, 48, 55, 78
organ, 55
outreach, 54
Overseas Private Investment Corporation, 94
oversight, ix, 2, 16
ozone, 33, 81, 82, 83, 84, 93

P

Pacific, 23, 54, 90, 94
pipeline, 46
plants, 25, 63
platform, ix, 11, 75
polar, 28
policy, viii, 1, 2, 14, 16, 19, 30, 31, 34, 38, 48, 49, 51, 56, 94
policymakers, 26
pollutants, 81, 82, 83, 84
pollution, 3, 45, 87
portfolio, ix, 22, 24, 47, 55
portfolio investment, 47
poverty, 51, 87
poverty reduction, 51
power generation, 23
predictability, 28, 30, 31
preparedness, ix, 22, 25, 63, 64, 87, 96
preservation, 50
President, vii, viii, ix, 1, 3, 4, 5, 6, 7, 9, 10, 11, 13, 14, 15, 16, 17, 19, 21, 22, 23, 24, 37, 42, 51, 53, 54, 56, 61, 62, 63, 64, 65, 75, 77, 83, 85
President Obama, 19, 23, 61, 77
principles, 25, 44
private investment, 94
private sector, ix, 11, 14, 15, 75, 78
procurement, 86
programming, viii, x, 2, 19, 49, 53, 56, 76, 79, 80, 81, 82, 83, 84, 92, 93, 94, 95
project, 22, 55, 60, 84, 86, 95
propane, 60
prosperity, 30, 33, 43
protection, 47, 81, 82, 83, 84
public health, 31, 47
public service, 85

Q

qualifications, vii, ix, 3
quality of life, 85
quantification, 28

R

race, 37
rainfall, 32
RE, 51
reading, 47
recommendations, ix, 2, 19, 23, 61
recovery, 60

reduce emissions, ix, 11, 43, 46, 49, 55, 75, 77
reform, 49, 86, 96
regulatory requirements, 47
regulatory systems, 51
reliability, 16
renewable energy, 42, 43, 45, 50, 51, 55, 57, 58, 59, 78
renewable fuel, 38
requirements, ix, 2, 16, 45, 46
research facilities, 31
resilience, ix, 22, 25, 56, 61, 63, 64, 77, 88
resolution, 28, 47, 84
resource allocation, 16
resource management, 8, 25, 29, 62, 78
resources, viii, 2, 8, 13, 14, 16, 19, 24, 25, 31, 32, 33, 46, 47, 61, 63, 64, 85, 87
response, 16, 19, 22, 31, 33, 43, 47, 50, 61, 87, 96
revenue, 27, 57, 59
risk(s), vii, 3, 4, 23, 25, 30, 31, 33, 48, 49, 51, 61, 64, 77, 87
risk management, 25
Rural Utilities Service, 39, 42, 66

S

safety, 46, 47
Saudi Arabia, 51
scarcity, 51
science, vii, viii, 1, 3, 5, 6, 8, 14, 18, 24, 25, 27, 28, 29, 30, 31, 32, 33, 46, 55, 64, 77
scientific knowledge, 30
scientific understanding, 28, 30, 33
scope, 5, 7, 15, 19
sea level, 15, 23, 32, 61
Secretary of the Treasury, 83
security, 23, 44, 46, 47, 48, 87
sellers, 51
Senate, x, 22, 76, 80
sensors, 28
services, 29, 30, 45, 51, 61, 77
shape, 78
Sierra Leone, 54
smog, 32
social sciences, 28
society, 33, 46
solution, 48, 87
South America, 91
South Asia, 91
Soviet Union, 54, 93
species, 34, 63, 64
spending, 6, 14, 20, 95
Spring, 95
stability, 4
stakeholders, 16, 63
state(s), vii, 3, 32, 34, 46, 55, 56, 61, 63, 73, 92, 94
storage, 37, 38
strategic management, 16
strategic planning, 29
structure, xi, 5, 57, 76
substitutes, 27, 59
survival, 33
sustainable economic growth, 23, 51

T

Tanzania, 54
target, 10, 22, 33, 77
Task Force, 23, 25, 61
tax credits, 27, 60
tax incentive, vii, viii, 2, 3, 24, 73
tax revenues, vii, viii, ix, 1, 3, 5
technical assistance, 32, 55
technical support, 64
techniques, 49
technological research, vii, 3
technologies, vii, 3, 8, 10, 14, 23, 24, 37, 38, 44, 45, 46, 47, 49, 50, 55, 94
technology, vii, 1, 3, 4, 5, 6, 14, 18, 19, 24, 37, 47, 50, 54
Tennessee Valley Authority, 11, 42, 43, 47, 72
tensions, 87
term plans, 29
terrestrial ecosystems, 31
the National Oceanic and Atmospheric Administration's, 94
threats, 87

time series, 7, 8
Title I, 19, 22, 81
Title IV, 19, 22
Title V, 84
trade, 51, 86
trade policy, 51
transaction costs, 45
transformation, 45
transmission, 24
transparency, 86
transportation, 32, 45, 46, 50
Treasury, x, 12, 27, 48, 50, 53, 54, 60, 61, 71, 75, 81, 82, 83, 84, 88, 95
trust fund, 78, 93
TVA, 47, 72

U

U.S. Department of Agriculture, 94
U.S. Department of Commerce, 94
U.S. Department of the Treasury, 48, 93
U.S. Geological Survey, 14, 28, 35, 69, 96
UN, 55, 89
United, ix, 3, 11, 21, 23, 57, 61, 75, 77, 86, 87, 93, 94, 95
United Nations, x, 12, 76, 94
United Nations Framework Convention on Climate Change, x, 12, 76, 94
United States, ix, 3, 11, 21, 23, 57, 61, 75, 77, 86, 87, 93, 94, 95
universe, 33
universities, vii, 3
USDA, 29, 30, 40, 42, 43, 65, 66, 67
USGS, 31, 62

V

vegetative cover, 43
vehicles, 19, 38, 45, 46, 57, 58, 59
vision, 47
vulnerability, 26, 48, 61, 64, 77, 88

W

Washington, 6, 7, 9, 11, 12, 19, 20
waste, 32
waste water, 32
water, 24, 25, 29, 30, 32, 33, 37, 45, 51, 53, 60, 63, 64, 77
water heater, 60
water quality, 32, 33, 63
water resources, 29, 30, 51, 63, 64
water supplies, 63
web, 54
welfare, 33
West Africa, 91
White House, 7, 8, 19, 94
wildfire, 30, 64
wildlife, 25, 63, 64
wind turbines, 60
windows, 60
workers, 21
workload, 16
World Bank, 78, 93, 96
worldwide, 31